Grade C Booster Workbook

Get your Grade C

NEW GCSE MATHS
Edexcel Linear

Matches the 2010 GCSE Specification

Greg Byrd

William Collins' dream of knowledge for all began with the publication of his first book in 1819. A self-educated mill worker, he not only enriched millions of lives, but also founded a flourishing publishing house. Today, staying true to this spirit, Collins books are packed with inspiration, innovation and practical expertise. They place you at the centre of a world of possibility and give you exactly what you need to explore it.

Collins. Freedom to teach.

Published by Collins
An imprint of HarperCollins*Publishers*
77–85 Fulham Palace Road
Hammersmith
London
W6 8JB

© HarperCollins*Publishers* Limited 2010

Browse the complete Collins catalogue at
www.collinseducation.com

10 9 8 7 6 5 4 3 2 1

ISBN-13 978-0-00-735382-8

Greg Byrd asserts his moral rights to be identified as the author of this work.

British Library Cataloguing in Publication Data
A Catalogue record for this publication is available from the British Library.

Commissioned by Katie Sergeant
Project managed by Lindsey Charles
Grade progression maps by Claire Powis, Maths AST; Children, Families and Education, Kent
Edited by Rosie Parrish
Answer check by Amanda Dickson
Cover design by Angela English
Concept design by Lesley Gray
Illustrations by Kathy Baxendale
Design and typesetting by Ken Vail Graphic Design
Production by Simon Moore
Printed and bound by Martins the Printers Ltd

Contents

Introduction

This workbook aims to help you get a C in your maths GCSE. It gives you plenty of practice in the key topics in the main sections of your course. These sections are colour coded: Number, Algebra, Geometry and Measures, Statistics and Probability.

Question grades

You can tell the grade of each question or question part by the colour of its number:

Grade B questions are shown as   (**1**)

Grade C questions are shown as   (**1**)

Grade D questions are shown as   (**1**)

Grade E questions are shown as   (**1**)

Higher level questions

Grades C and D questions occur on both Foundation and Higher papers but some topics are only on Higher papers. These topics are shown by this flag. If you are taking the Foundation papers, you can ignore these topics.

> ! Higher
> • Tier only

Remember...

These boxes remind you about things which students often forget or get wrong. Take notice of them! If you are taking exams in 2011 or later there is a small amount of material you do not need to know – it will be marked by a box like this.

> **Remember:** Significant figures and decimal places are not the same.

Exam practice

The revision sections help you to prepare for the exams. There are two, one to help you prepare for each exam paper. In the first revision section you are not allowed to use a calculator (as in paper 1) and in the second you can use a calculator (as in paper 2).

Grade progression maps

At the back of the book are grade progression maps for the main subject areas. These show how you can move from a grade D to a grade C in each one. Use them to check what you know and can do before the exam!

Answers

Finally there are answers to all the questions at the back of the book. You can check your answers yourself or your teacher might tear them out and give them to you later to mark your work.

Rounding / Giving an approximate answer

1 Round each of these numbers to one significant figure.

    a 5.7 = _____   **b** 0.0626 = _____   **c** 751.3 = _____

Remember: Significant figures and decimal places are not the same.

2 Round each of these numbers to two significant figures.

    a 22.475 = _____   **b** 0.868 = _____   **c** 57392.1 = _____

3 Round each of these numbers to three significant figures.

    a 91.042 = _____   **b** 0.037885 = _____   **c** 7896.4833 = _____

4 Round each of the following to the number of significant figures (sf) indicated.

    a 3750 (1 sf) = _____   **b** 7.3861 (2 sf) = _____   **c** 0.005555 (3 sf) _____

    d 99.8933 (3 sf) = _____   **e** 99.8933 (2 sf) = _____   **f** 99.8933 (1 sf) = _____

5 Use your calculator to work out the following. Give your answers correct to three significant figures.

    a $\dfrac{185^2}{0.47}$ = _____   **b** $\dfrac{989 \times 0.618}{320 \times 0.053}$ _____   **c** $\dfrac{599}{0.41} + \dfrac{209}{0.51}$ = _____

6 Use approximations to estimate an answer to each of the following. You must show your workings.

    a 72.3×18.4 **b** $193.42 \div 38.09$

    c $\dfrac{31.9 \times 13.9}{1.18 - 0.52}$ **d** $\dfrac{8.77 \times 5.08}{19.3 - 4.9}$

Rounding / Giving an approximate answer

Remember: When approximating with square roots, round the number under the root to a square number.

e $\sqrt{(21.93 \times 3.84)}$

f $\dfrac{185^2}{0.47}$

g $\dfrac{989 \times 0.618}{320 \times 0.053}$

h $\dfrac{599}{0.41} + \dfrac{209}{0.51}$

i $\dfrac{\sqrt{(69.96 \div 2.11) \times 18.75}}{9.81 \div 0.48}$

j $\sqrt{0.038} - \sqrt{0.38} - \sqrt{0.8}$

Prime factors / Lowest common multiples / Highest common factors

1 a Write 28 as the product of its prime factors. Give your answer in index form.

b Write 72 as a product of its prime factors. Give your answer in index form.

_____ _____

2 Find the lowest common multiple (LCM) of

a 5 and 7

b 28 and 72

_____ _____

3 Find the highest common factor (HCF) of

a 12 and 20

b 28 and 72

_____ _____

4 Tariq coughs every two minutes.

Henry sneezes every five minutes.

Paul moans every eight minutes.

Tariq, Henry and Paul cough, sneeze and moan at the same time.

In how many minutes will they next cough, sneeze and moan together at the same time?

_____

Indices

1 Write these expressions using power notation.

a $3 \times 3 \times 3 \times 3 \times 3 = $ _____

b $10 \times 10 \times 10 = $ _____

c $2 \times 2 \times 2 \times 2 \times 2 \times 2 = $ _____

2 Write these expressions using power notation.

a $a \times a \times a \times a = $ _____

b $b \times b = $ _____

c $x \times x \times x \times x \times x \times x \times x \times x = $ _____

3 Simplify each expression. Give each answer as a number or letter to a single power.

a $5^2 \times 5^2 = $ _____

b $7^3 \times 7^5 = $ _____

c $2^{17} \times 2^3 = $ _____

d $3^9 \div 3^3 = $ _____

e $6^5 \div 6^4 = $ _____

f $8^{20} \div 8^{15} = $ _____

g $y^2 \times y^2 = $ _____

h $b^3 \times b^5 = $ _____

i $j^{17} \times j^3 = $ _____

j $m^6 \div m^2 = $ _____

k $a^9 \div a^8 = $ _____

l $t^4 \div t^4 = $ _____

m $\dfrac{7^6}{7^4} = $ _____

n $\dfrac{x^5}{x^2} = $ _____

o $\dfrac{x^2}{x^5} = $ _____

p $d^2 \times d^3 \times d^4 = $ _____

q $2a^2 \times 2a^3 \times a^4 = $ _____

r $h^4 \times h^5 \div h^3 = $ _____

s $\dfrac{x^3 \times 3x^2}{x} = $ _____

t $\dfrac{b^2 \times 6b^3}{2b^4} = $ _____

u $\dfrac{4t^3 \times 3t^5}{6t^4 \times t} = $ _____

Fractions

(1) Work out each of the following. Give your answers in their simplest form and as mixed numbers when possible.

a $\frac{3}{4} + \frac{1}{7}$ = _____

b $\frac{2}{5} + \frac{4}{9}$ = _____

c $3\frac{1}{3} + 1\frac{3}{5}$ = _____

d $\frac{3}{8} + \frac{3}{10} + \frac{4}{5}$ = _____

e $\frac{5}{6} - \frac{3}{5}$ = _____

f $\frac{8}{9} - \frac{3}{4}$ = _____

g $4\frac{1}{4} + 2\frac{2}{3}$ = _____

h $2\frac{2}{3} + 1\frac{3}{4}$ = _____

i $5\frac{1}{2} - \frac{5}{8} + 1\frac{3}{5}$ = _____

j $8 - 1\frac{3}{4} - \frac{7}{8}$ = _____

k $\frac{3}{4} \times \frac{1}{2}$ = _____

l $\frac{8}{9} \times \frac{3}{4}$ = _____

m $1\frac{3}{4} \times 2\frac{1}{2}$ = _____

n $2\frac{1}{2} \times 3\frac{1}{5}$ = _____

o $4 \times 2\frac{5}{8}$ = _____

p $8\frac{2}{5} \times 4$ = _____

q $\frac{1}{4} \div \frac{1}{3}$ = _____

r $\frac{1}{3} \div \frac{1}{4}$ = _____

s $6\frac{2}{5} \times 1\frac{3}{5}$ = _____

t $1\frac{3}{5} \div 6\frac{2}{5}$ = _____

(2) A patio is $3\frac{1}{4}$ metres long and $2\frac{1}{5}$ metres wide. What is the area of the patio?

_____

(3) Amy's stride is four-fifths of a metre long. How many strides does she take to walk the length of a 12 m long bus?

_____

Percentage increase and decrease

1 Increase £250 by 12%.

_____

2 Increase £2.50 by 12%.

_____

3 Decrease 45 m by 12%.

_____

4 Decrease 4.5 m by 12%.

_____

5 Last year Xavier's salary was €21 550. This year it is €22 412.
What is the percentage increase of Xavier's salary?

_____

6 A van is for sale for £12 500 + VAT (at 17.5%).
Work out the total cost of the van.

_____

7 ABCD is a rectangle with
length 35 cm and width 10 cm.
The length of the rectangle is increased by 10%.
The width of the rectangle is decreased by 10%.

```
A                                    B
 ┌──────────────────────────────────┐
 │                                   │ 10 cm
 │                                   │
 └──────────────────────────────────┘
C              35 cm                 D
```

a Has the area of the rectangle increased or decreased?

_____

_____

b What is the actual increase or decrease in the area of the rectangle?

_____

8 A flat screen TV usually sells for £395 + VAT (at 17.5%). It is on
sale for 20% off.

What is the new cost of the TV?

> **Remember:** Do not simply
> take 2.5% off the £395.

_____

_____

One quantity as a percentage of another

1 Express the first quantity as a percentage of the second.

    a £5, £25 = _____   **b** 8 kg, 32 kg = _____

    c 3.2 ml, 6 ml = _____   **d** 12 minutes, 2 hours = _____

    e 375 g, 1 kg = _____   **f** 2 hrs, 1 day = _____

2 In her English test, Jodie scored 24 out of 40. In her History test she scored 56 out of 90. By using percentages, show that Jodie is better at History than English.

_____

3 Harri bought a car for £1500 and sold it for £1800. What is Harri's percentage profit?

_____

4 Before Dec started his diet, he weighed 85 kg. He now weighs 77 kg. What percentage of his original weight has he lost?

_____

5 In the 2008/2009 football season, West Bromwich Albion came bottom of the premier league. They won eight of their 38 league games. What percentage of games did they *not* win?

_____

6 In the 2007/2008 football season, West Bromwich Albion came top of division 1. They lost eleven of their 46 league games. What percentage of games did they *not* lose?

_____

Compound interest

1 Increase £70 by 5%.

2 Decrease £5 by 70%.

3 Philippe has $2000 to invest. What does the calculation $\$2000 \times 1.05^3$ represent?

4 £1000 is invested in an account that pays 4.5% compound interest per annum. How much will the account be worth after four years?

5 The islands of the Maldives had a population of 390 000 at the start of 2009. Their population growth rate is 5.6% per annum. Estimate the size of the population at the start of 2015.

6 A certain type of conifer hedge grows at a rate of 15% each year for the first 18 years. Mr Jones buys a one year old plant that is 50 cm tall.

    a How tall will it be in 15 years' time?

    b How long did it take it to grow to 3 m?

7 A £20 000 car depreciates by 12% a year. How much is it worth after five years?

8 Although some experts in 2009 put numbers of hedgehogs as high as one million, hedgehogs are on the endangered species list as their numbers continue to fall by around 5% a year. How many hedgehogs are there likely to be in 10 years' time?

Sharing an amount by a given ratio

1 Share 50 in the ratio 2:3.

2 Share 400 g in the ratio 2:3.

3 Share £90 in the ratio 3:7.

4 Share 1 km in the ratio 1:99.

5 Share 100 ml in the ratio 2:3:5.

6 Share 13.5 kg in the ratio of 6:5:1.

7 Concrete is made from cement, sand and gravel in the ratio of 1:2:4.
Paul needs three and a half tonnes of concrete. How much of each ingredient does he need?

8 Aaron, Beth and Charlie are going to share £5940 in the same ratio as their ages.
Aaron, Beth and Charlie are 6, 9 and 15 years old respectively.

a How much do they each receive?

b If the money was shared out one year later, who would benefit the most, and by how much?

9 The sizes of the interior angles of a quadrilateral are in the ratio 2:3:6:7.
Calculate the size of the smallest angle.

Ratio

1 The ratio of male teachers to female teachers in a primary school is 1:9. There are 36 female teachers in the school. How many male teachers are there?

2 The ratio of male teachers to female teachers in a primary school is 8:7. There are 32 male teachers in the school. How many teachers are there altogether?

3 The ratio of staff to pupils in a school is 2:35. There are 665 pupils in the school. How many staff are there?

4 The ratio of staff to year 10 students to year 11 students on a school trip is 2:5:7. There are 225 year 10 students on the trip.

a How many year 11 students are on the trip?

b How many people are on the trip altogether?

5 Mahmoud is making a drink from lemonade, orange and ginger in the ratio 40:9:1. He has 4.5 litres of orange and uses it all. How much of the other ingredients does he use?

6 When the cost of a meal was shared between two families in the ratio of 4:5, the smaller share was £31. How much did the meal cost altogether?

Basic powers and roots

1 Complete this sequence of square numbers from 1^2 to 15^2.

1, 4, 9, _____

2 Complete this sequence of cube numbers from 1^3 to 5^3.

1, 8, _____

3 Without a calculator, work out the following.

a $\sqrt{100} =$ _____

b $\sqrt{144} =$ _____

c $\sqrt[3]{64} =$ _____

d $\sqrt[3]{1} =$ _____

e $\sqrt[3]{1000} =$ _____

f $\sqrt{49} + \sqrt[3]{125} =$ _____

g $\sqrt[3]{5^2 + 4^2 - 3^2 - 2^2 - 1^2} =$ _____

4 Write down *both* square roots of 16: _____ and _____

5 Explain why −10 is one of the square roots of 100.

_____

6 Use your calculator to work out the following.

a $\sqrt{676} =$ _____

b $17.4^2 =$ _____

c $\sqrt[3]{29.791} =$ _____

d $0.7^3 =$ _____

e $\sqrt{0.04^3} =$ _____

f $\sqrt[3]{64^2} =$ _____

7 Without a calculator, work out the following.

a $\dfrac{\sqrt{13^2 - 5^2}}{2^3 - 2} =$

_____

b $\dfrac{4^3 - 2^2}{\sqrt{5^3 - 5^2}} =$

_____

8 When $x = 2$ and $y = 5$, find the value of the following, without using a calculator.

a $x^2 + y^2 =$ _____

b $y^3 - x^3 =$ _____

c $\sqrt{y^2 + x^2 + y^2 - y} =$ _____

Reciprocals

1 Without using a calculator, find the reciprocal of each of the following integers. Give your answers as (i) a fraction and (ii) a decimal.

a 2 = (i) _____ = (ii) _____   **b** 5 = (i) _____ = (ii) _____

c 10 = (i) _____ = (ii) _____   **d** 100 = (i) _____ = (ii) _____

2 Look at your answers to question 1. What do you notice about what happens to the size of the decimal answers as the size of the integers gets bigger?

_____

3 Use your calculator to find the reciprocal of each of the following decimal numbers.

a 0.25 = _____   **b** 2.5 = _____   **c** 25.0 = _____   **d** 0.025 = _____

4 Find the reciprocals of each of the following fractions. Give your answers as mixed numbers where appropriate.

a $\frac{7}{9}$ = _____   **b** $1\frac{7}{9}$ = _____   **c** $2\frac{7}{9}$ = _____   **d** $3\frac{7}{9}$ = _____

5 **a** What is the reciprocal of 0.4? _____

b What is the reciprocal of your answer to part **a**? _____

c Multiply 0.4 by its reciprocal. _____

d Multiply 0.5 by its reciprocal. _____

e Look at your answers to parts **c** and **d**. Does this always happen?

_____

6 Johan says that the reciprocal of a number is always smaller than the number. Give an example to show that Johan is wrong.

_____

Limits

! Higher
• Tier only

Remember: You should use either ≤ and < or ≤ and ≤.

1 Write down the limits of accuracy of the following.

a 5 cm measured to the nearest centimetre = _____

b 50 km/h measured to the nearest 10 km/h = _____

c 50 students measured to the nearest 10 students = _____

d £50 measured to the nearest £10 = _____

e 300 m/s measured to the nearest 10 m/s = _____

f 300 puppies measured to the nearest 100 puppies = _____

2 Write down the limits of accuracy of the following.

a 25 cm measured to the nearest centimetre = _____

b 25 cm measured to the nearest 5 cm = _____

c 47 500 football fans measured to the nearest 10 fans = _____

d 12.7 cm measured to the nearest 0.1 cm = _____

3 There are 500 sweets in a jar, measured to the nearest 10. The total mass of the sweets is 500 g to the nearest 10 g.

Remember: Bounds and limits are the same thing.

a Between what bounds does the number of sweets lie? _____

b Between what bounds does the mass of the sweets lie? _____

c Why are the answers to parts **a** and **b** different? _____

_____

4 A poster is a rectangle with a length of 40 cm and a width of 80 cm. Both measurements are correct to the nearest centimetre.

a What is the greatest possible length of the poster? _____ cm

b What is the least possible width of the poster? _____ cm

c What is the greatest possible area of the poster? _____ cm^2

Factorising

1 Factorise the following.

a $6a + 12 =$ _____

b $4a + 8b =$ _____

c $4x + 6y =$ _____

d $8t - 6p =$ _____

e $2ab + 6ac =$ _____

f $5mn - 5mp =$ _____

g $p^2 + 5p =$ _____

h $7h - h^2 =$ _____

i $3x^2 + 2x =$ _____

j $3t^2 - 3tp =$ _____

k $6x^2 + 9xy =$ _____

l $12a^2 - 8ab =$ _____

m $4b^2c + 8bc =$ _____

n $8abc - 6bed =$ _____

o $2ab + 4a^2b =$ _____

p $4x^2 + 6x + 8y =$ _____

q $6mp + 9bm + 3mt =$ _____

r $8cd^2 - 2cd - 4c^2d - 12\ c^2d^2 =$ _____

2 **a** Factorise $n^2 - n =$ _____

b When n is a whole number, explain why $n^2 - n$ is always an even number.

_____

_____

3 Write down an expression for the missing lengths of each of the following rectangles.

a
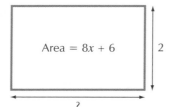
Area = $8x + 6$ 2

?

length = _____

b
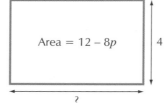
Area = $12 - 8p$ 4

?

length = _____

c
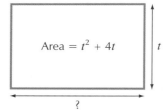
Area = $t^2 + 4t$ t

?

length = _____

4 Look at these shapes.

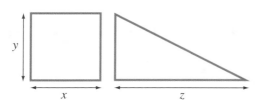

An expression for the total area of the square and the triangle is $xy + \frac{1}{2}zy$.
Factorise the expression for the total area of the square and the triangle.

_____

Brackets

Remember: Always be very careful about negative numbers.

1 Expand the following expressions.

    a $3(a + 2) =$ _____   **b** $3(4 + x) =$ _____   **c** $3(x - y) =$ _____

    d $x(x + 1) =$ _____   **e** $x(7 - x) =$ _____   **f** $x(x - y) =$ _____

    g $3a(3 + a) =$ _____   **h** $3a(3a + b) =$ _____   **i** $3a^2(4a - 3b) =$ _____

2 Expand and simplify the following expressions.

    a $2(x - 3) + 3(x + 6) =$ _____

    b $5(2x + y) + 4(2y + x) =$ _____

    c $x(x + 3) + x(2x - 1) =$ _____

    d $3x(x + 3) + 2x(2x - 1) =$ _____

    e $4a(2a + 3b) - 3a(2a + 2b) =$ _____

    f $3x(2x + 3y) + 2y(3y - 2x) =$ _____

    g $5a(3a + 4) - 2a(3 - 4a) =$ _____

    h $5p(2q - 2r) + 2q(3p - r) =$ _____

3 Show that

    a $3(a + 2) + 9(a + 1) = 3(4a + 5)$ **b** $2(4b + 1) - 4(b - 3) = 4(b + 4) - 2$

4   **a** Write an expression for the total area of both rectangles using brackets.

    b Expand and simplify your answer to part **a**.

Double brackets

Remember: Always be very careful with negative numbers.

1 Expand and simplify the following expressions.

a $(x + 2)(x + 3) =$ _____

b $(p + 4)(p + 2) =$ _____

c $(a + 4)(a - 2) =$ _____

d $(d - 3)(d + 5) =$ _____

e $(p - 4)(p - 5) =$ _____

f $(3 + x)(5 + x) =$ _____

g $(5 + y)(2 - y) =$ _____

h $(x + 1)(x - 1) =$ _____

i $(q + 10)(q - 10) =$ _____

j $(x + 4)^2 =$ _____

k $(4 + x)^2 =$ _____

l $(a - 2)^2 =$ _____

2 Show that $(x + 5)(x - 3) - 2x = x^2 - 15$

_____

_____

3 **a** Write an expression for the area of the rectangle, using brackets.

_____

b Expand and simplify your answer to part **a**.

_____

$x - 3$

$x + 2$

4 **a** Write an expression for the area of the rectangle, using brackets.

_____

b Expand and simplify your answer to part **a**.

_____

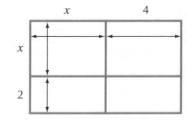

ALGEBRA

16

Solving linear equations

1 Solve the following equations.

a $4x - 3 = 5$

b $3 - 4x = 1$

c $\dfrac{x}{3} + 7 = 10$

d $2(3x + 1) = 32$

_____

_____

_____

_____

2 Solve these equations.

a $\dfrac{12 - x}{3} = 5$

b $\dfrac{17 - x}{3} = 4.5$

c $\dfrac{20 - 2x}{4} = 3$

_____

_____

_____

d $5x + 13 = 3(x + 5)$

e $2(1 - 5x) = 3(5x - 1)$

_____

_____

f $3(x - 3) = 7(x - 2) - 3$

g $3(2x - 1) + 4(x + 3) = 5(2x - 1) + 4(3x - 1)$

_____

_____

Set up and solve linear equations

1 An isosceles triangle has angles of $2x°$, $2x°$ and $6x°$.

    a Form an equation in terms of x. _____

    b Solve your equation to find the value of x. _____

    c Give the sizes of the angles of the isosceles triangle. _____

2 Tom thought of a number. He divided it by two, and then added four. The result is one more than the number he first thought of.

    a Use the above information to set up an equation using n to represent Tom's number.

      _____

    b Solve your equation to find the value of n. _____

      _____

3 Adam drew two rectangles with the same area.

    5   [rectangle]   3   [rectangle] $x + 4$

    $x - 2$

    a Set up an equation in terms of x. _____

    b Solve your equation to find the value of x. _____

      _____

      _____

      _____

4 Hammy used a computer to draw a regular hexagon and a regular octagon, both with the same perimeter. The side length of the hexagon is $2x - 1$. The side length of the octagon is $x + 5$.

    a Set up an equation in terms of x. _____

    b Solve your equation to find the value of x. _____

      _____

      _____

5 Matt draws a square of side length $2x$.
He then draws a triangle with side lengths
of $2x + 5$, $3x - 1$ and $4x - 7$. The triangle and the
square have the same perimeter.
Find the value of x.

6 Graham is G years old. His son is 25 years younger
than he is. The sum of their ages is 41.
How old is Graham?

7 Kate is K years old. Her cousin is twice as old as
she is. The sum of their ages is 36.
How old is Kate?

8 Paul thought of a number. He multiplied his number
by four. He added three to the result. He doubled that
result and got 54.
What number did he start with?

9 Lilly starts with £21.50. Holly starts with £17.20. Lilly
buys five coloured phone covers and Holly buys three.
The phone covers are all the same price. Lilly and
Holly have the same amounts of money left over.
Find the price of one of the phone covers.

Trial and improvement

1 For each of the following equations, find a pair of consecutive whole numbers between which the solution lies.

Remember: Consecutive is when one number follows another e.g. 4, 5, 6 are consecutive numbers.

a $x^2 - x = 28$ **b** $x^2 + x = 28$ **c** $x^3 + 2x = 85$

_____   _____   _____

2 Use trial and improvement to find a solution to the equation $x^3 - 2x = 71$, correct to one decimal place.

x	$x^3 - 2x$	comment
3	21	too small
6	204	too big

Answer: _____

3 Use trial and improvement to find a solution to the equation $\dfrac{x^3}{2} + x = 93$, correct to one decimal place.

x	$\dfrac{x^3}{2} + x = 93$	comment
3	16.5	too small
7	17.5	too big

Answer: _____

4 The area of this rectangle is $750\,\text{cm}^3$.

$(x^2 + 10)$ cm

x cm

a Write an equation showing this information.

_____

b Use trial and improvement to find a solution to the equation, correct to one decimal place.

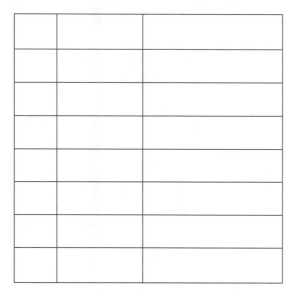

Answer: _____

5 Use trial and improvement to find a solution to the equation $2x^2 + \sqrt{x} = 17$, correct to one decimal place.

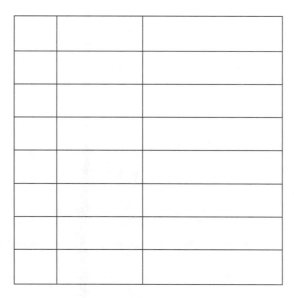

Answer: _____

Rearranging (changing the subject of) formulae

1 Rearrange these formulae to make a the subject.

a $x = a + 6$ **b** $y = a - 6$ **c** $z = 6 - a$

_____   _____   _____

_____   _____   _____

2 Rearrange these formulae to make b the subject.

a $x = 4b + 6$ **b** $y = 5b - 3$ **c** $z = 6 - 2b$

_____   _____   _____

_____   _____   _____

3 Rearrange these formulae, the subject is given in brackets.

a $V = IR$ (R) **b** $P = 2b + 2b$ (b) **c** $A = \dfrac{bh}{2}$ (b)

_____   _____   _____

_____   _____   _____

d $y = mx + c$ (c) **e** $y = mx + c$ (m) **f** $v = u + at$ (t)

_____   _____   _____

_____   _____   _____

g $A = \pi r^2$ (π) **h** $F = \dfrac{9C}{5} + 32$ (C) **i** $A = \pi ab$ (a)

_____   _____   _____

_____   _____   _____

4 Rearrange these formulae, the subject is given in brackets.

a $A = \pi r^2$ (r) **b** $V = \pi r^2 h$ (r) **c** $V = \dfrac{4\pi r^3}{3}$ (r)

_____   _____   _____

_____   _____   _____

_____   _____   _____

ALGEBRA

The nth term

1 Find the nth term of each sequence.

    a 5, 10, 15, 20, … = _____

    b 2, 4, 6, 8, … = _____

    c 50, 100, 150, 200, … = _____

    d 20, 40, 60, 80, … = _____

2 Find the nth term of each sequence.

    a 6, 11, 16, 21, … = _____

    b 3, 5, 7, 9, … = _____

    c 3, 7, 11, 15, … = _____

    d 1, 6, 11, 16, … = _____

    e −4, −1, 2, 5, 8, … = _____

    f −80, −70, −60, −50, … = _____

3 Find the nth term of each sequence.

    a 10, 7, 4, 1, … = _____

    b 20, 16, 12, 8, … = _____

    c 97, 87, 77, 67, … = _____

    d −5, −10, −15, −20, … = _____

    e 0, −3, −6, −9, … = _____

    f 1, −9, −19, −29, … = _____

4 Find the nth term of each sequence.

    a $\dfrac{3}{4}, \dfrac{5}{9}, \dfrac{7}{14}, \dfrac{9}{19}, \ldots =$ _____

    b $\dfrac{1}{2}, \dfrac{2}{5}, \dfrac{3}{8}, \dfrac{4}{11}, \ldots =$ _____

    c $\dfrac{3}{97}, \dfrac{7}{87}, \dfrac{11}{77}, \dfrac{15}{67}, \ldots =$ _____

    d $\dfrac{-1}{6}, \dfrac{-5}{7}, \dfrac{-9}{8}, \dfrac{-13}{9}, \ldots =$ _____

5 Look at this sequence of squares.

  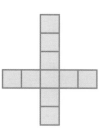 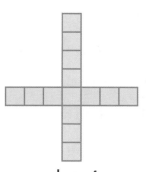

  shape 1   **shape 2**   **shape 3**   **shape 4**

Write down the number of squares used in shape n.

_____

The nth term

6 A catering company uses tables in the shape of a trapezium.
The diagrams below show the number of people that can sit at different numbers of tables.

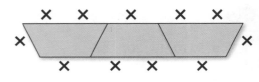

    1 table   **2 tables**   **3 tables**

    a In this arrangement, how many people could sit at n tables? _____

    b For a charity event, up to 150 people have to be seated. How many tables arranged like this do they need?

    _____

    _____

    _____

    _____

7 Find the nth term of each sequence.

    a 2, 4, 8, 16, 32, … = _____
        b 10, 100, 1000, 10 000, … = _____

8 The nth term of a sequence is $5n + 2$.

    a Write down the first four terms of the sequence. _____

    b Write down the 100th term of the sequence. _____

9 The nth term of a sequence is $\frac{1}{2}n(n + 1)$.

    a Write down the first four terms of the sequence. _____

    b Write down the 100th term of the sequence. _____

10 The nth term of a sequence is $n^2 + n$.

    a Write down the first four terms of the sequence. _____

    b Write down the 100th term of the sequence. _____

11 The nth term of a sequence is $2n^2 - n - 1$.

    a Write down the first four terms of the sequence. _____

    b Write down the 100th term of the sequence. _____

Inequalities

1 Solve the following inequalities.

a $x + 3 < 8$

_____

b $x - 2 \geq 12$

_____

c $x - 6 \leq 1$

_____

d $3x + 5 < 17$

_____

_____

e $2x + 3 \geq 5$

_____

_____

f $5x - 2 \leq -12$

_____

_____

g $\frac{x}{5} > 6$

_____

_____

h $\frac{x}{2} + 1 \geq 5$

_____

_____

i $3(x + 4) < 6$

_____

_____

2 Write down the largest integer value of x that satisfies each of these inequalities.

a $x + 3 \leq 8$

_____

$x =$ _____

b $x - 2 < 12$

_____

$x =$ _____

c $x + 7 < 3$

_____

$x =$ _____

d $3x + 5 < 18$

_____

_____

_____

$x =$ _____

e $3(x + 4) \leq -4$

_____

_____

_____

$x =$ _____

f $5x - 3 < 12$

_____

_____

_____

$x =$ _____

3 Write down the largest integer value of x that satisfies each of these inequalities.

a $2x + 3 < 22$, where x is a square number.

_____

_____

$x =$ _____

b $2x + 1 \leq 30$, where x is a prime number.

_____

_____

$x =$ _____

Inequalities

4 What inequalities are shown by the following number lines?

a

b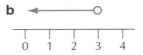

c

d

a _____   b _____   c _____   d _____

5 **a** Which integers satisfy both inequalities in questions **4a** and **4b** above? _____

b Which integers satisfy both inequalities in questions **4b** and **4c** above? _____

c Which integers satisfy both inequalities in questions **4c** and **4d** above? _____

6 Draw the following inequalities on a number line.

a $x \geq 2$

b $x > 3$

c $x \leq 8$

d $2 \leq x \leq 4$

e $-3 < x < 0$

f $7 \leq x < 10$

7 Solve the inequality $2(4x + 3) \leq 18$ and illustrate its solution on the number line.

_____

_____

_____

8 ❗ Higher
   Tier only

a Draw the line $x = -3$ on the axes.

b Draw the line $x = 1$ on the axes.

c Shade the region defined by $-3 \leq x \leq 1$.

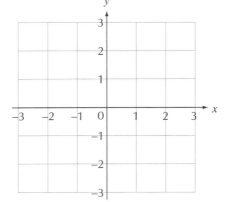

ALGEBRA

26

Real-life graphs

1 Dom was travelling in his car to the airport. He set off from home at 10:00 am, and stopped on the way for a break. This distance-time graph illustrates his journey.

a At what time did he

    i stop for his break? _____

    ii set off after his break? _____

    iii get to the airport? _____

b What was his average speed

    i during the first hour? _____

    ii during the last part of his journey? _____

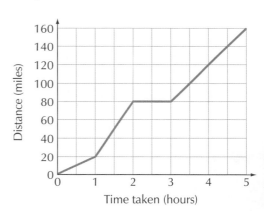

2 Work out the average speed of each of the journeys shown by these graphs.

a

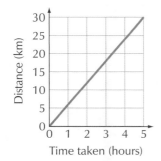

Speed = _____ km/h

b

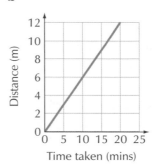

Speed = _____ m/min

c

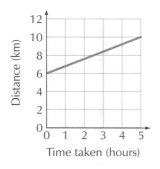

Speed = _____ km/h

d

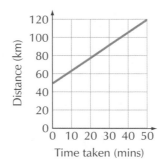

Speed = _____ km/min

Real-life graphs

3 **a** Calculate the average speed during each stage of the journey shown on the graph.

A to B = _____ km/h

B to C = _____ km/h

C to D = _____ km/h

D to E = _____ km/h

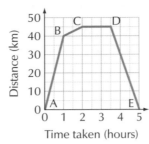

b By looking at the graph, how can you tell which part of the graph shows the fastest speed?

_____

4 The graph shows a race between Rob and Darren. Describe what happens in the race.

_____

_____

_____

_____

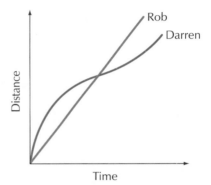

5 Water is poured at a steady rate into these jars.

A 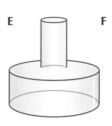 B  C  D 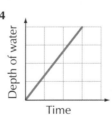 E 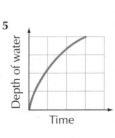 F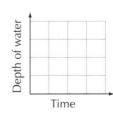

The depth of the water in the jars is measured over time and graphs are drawn.

1 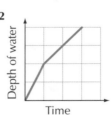 2 3 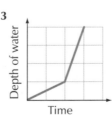 4 5

a Match the jars to the graphs: A = _____ B = _____ C = _____ D = _____ E = _____ F = _____

b One jar has not been matched. Sketch a graph for this jar.

Gradient and intercept

1 Work out the gradients of lines A to F.

A = _____

B = _____

C = _____

D = _____

E = _____

F = _____

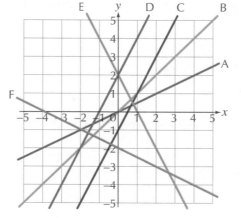

2 Write down the gradient of each of these lines.

a $y = 4x + 3$

gradient = _____

b $y = 3 + x$

gradient = _____

c $y = -\frac{1}{2}x + 5$

gradient = _____

d $y = 6 - 2x$

gradient = _____

3 **a** Which line in question 2 is the steepest? _____

b How do you know? _____

4 Write down the y-axis intercepts of each of these lines.

a $y = 3x + 4$

intercept = _____

b $y = 3 + 2x$

intercept = _____

c $y = -\frac{x}{3} + 2$

intercept = _____

d $y = 3x$

intercept = _____

5 Look at the straight line equations.

A: $y = x + 2$ B: $y = 2x + 1$ C: $y = 2x + 2$ D: $y = 3x - 1$ E: $y = x$

F: $y = -x$ G: $3 - x = y$ H: $y = 3 - 2x$ I: $y = 4 - 2x$

a Write the letter of the line with the steepest gradient. _____

b Which of the lines are parallel to each other? _____

c Write the letter of the line that crosses the y-axis at the highest point. _____

d Write the letter of the line that crosses the y-axis at the lowest point. _____

e How many of the lines have a negative gradient? _____

Drawing graphs – linear

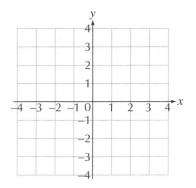

1 On the coordinate grid opposite draw and label these lines.

a $x = 2$ **b** $y = -3$

c $x = -3$ **d** $y = 2$

2 a Complete this table of values for $y = 2x - 1$ for $-2 \leq x \leq 2$.

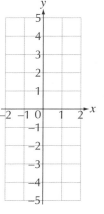

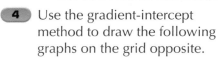

x	–2	–1	0	1	2
y	–5			1	

b Draw the graph of $y = 2x - 1$ on the grid opposite.

c Use the graph to find the value of x when $y = 0$.

$x = $ _____

3 a Draw the graph of $y = 3x + 2$ for $-3 \leq x \leq 3$.

b Use the graph to find the value of x when $y = -4$.

$x = $ _____

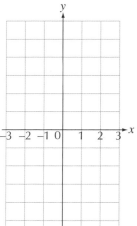

> **Remember:** The gradient-intercept method means you start by plotting the intercept and then plot other points by using the gradient.

4 Use the gradient-intercept method to draw the following graphs on the grid opposite.

a $y = 2x + 2$

b $y = -\dfrac{1}{2}x + 2$

c $y = x + 2$

d $y = 5 - 2x$

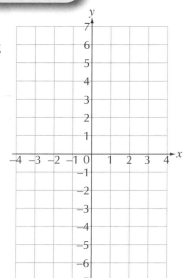

ALGEBRA

Drawing graphs – quadratic

1 **a** Complete this table of values for $y = x^2 + 2$ for $-4 \leq x \leq 3$.

x	−4	−3	−2	−1	0	1	2	3
y	18			3			6	

b Draw the graph of $y = x^2 + 2$ on the grid opposite.

c Use the graph to find the value of y when $x = 1.5$.

$y =$ _____

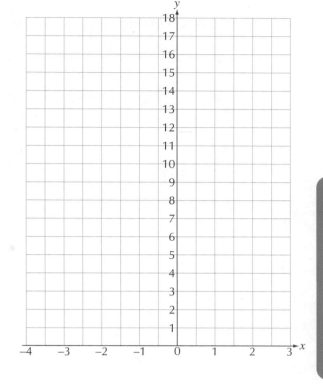

2 **a** Complete this table of values for $y = 2x^2 - 4x - 1$.

x	−2	−1	0	1	2	3
y		5	−1		−1	

b Draw the graph of $y = 2x^2 - 4x - 1$ on the grid opposite.

c An approximate solution of the equation $y = 2x^2 - 4x - 1$ is −0.22

  i Explain how you can find this from your graph.

  _____

  ii Use your graph to write down another solution of this equation.

  _____

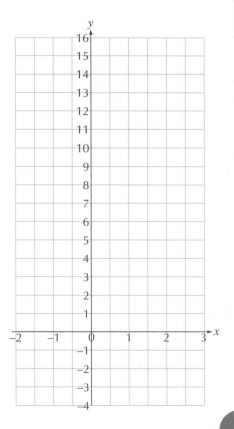

Circles – circumference

1 Calculate the circumference of each of the following circles.

Give all answers correct to one decimal place.

a
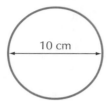
10 cm

_____

_____

_____

b
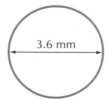
3.6 mm

_____

_____

_____

c
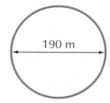
190 m

_____

_____

_____

d
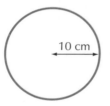
10 cm

_____

_____

_____

e
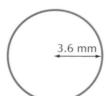
3.6 mm

_____

_____

_____

f
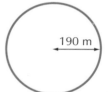
190 m

_____

_____

_____

2 Calculate the circumference of each of the following circles.

Leave all answers in terms of π.

a
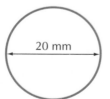
20 mm

_____

_____

_____

b
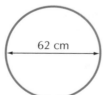
62 cm

_____

_____

_____

c
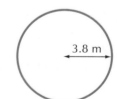
3.8 m

_____

_____

_____

3 A bicycle wheel has a diameter of 65 cm.

    a Work out the circumference of the wheel.

    _____

    _____

    b How many complete revolutions does the wheel make when the bicycle travels 500 m?

    _____

    c The 'Tour de France' is approximately 3500 km long. How many complete revolutions does the wheel make when the bicycle completes the 'Tour de France'?

    _____

    _____

    _____

4 A penny-farthing bicycle has a front wheel with a radius of 90 cm and a back wheel with a radius of 20 cm.

In a journey the front wheel turned 200 times.

> **Remember:** Find the distance travelled first.

How many complete revolutions did the rear wheel make?

_____

_____

_____

5 Calculate the perimeter of the following shapes.
Give your answers correct to one decimal place.

> **Remember:** Include the lengths of any straight sides.

a

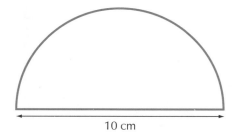

10 cm

b

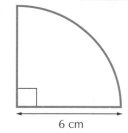

6 cm

_____   _____

_____   _____

_____   _____

GEOMETRY AND MEASURES

Circles – circumference

6 Calculate the perimeter of the following shapes.
Give your answers in terms of π.

a

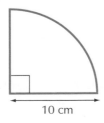

10 cm

b

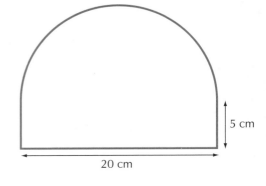

5 cm

20 cm

_____

_____

_____

_____

_____

_____

7 A circle has a circumference of 300 cm.
Calculate the diameter of the circle to one decimal place.

_____

_____

8 A circle has a circumference of 1 m.
Calculate the radius of the circle to two decimal places.

_____

_____

9 Find the total length of the black lines in this shape made of two circles and three straight lines.

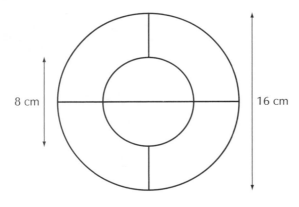

8 cm

16 cm

_____

_____

_____

GEOMETRY AND MEASURES

Circles – area

1 Calculate the area of each of the following circles.
Give all answers correct to one decimal place.

a

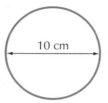
10 cm

_____

_____

_____

b

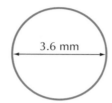
3.6 mm

_____

_____

_____

c

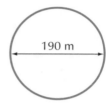
190 m

_____

_____

_____

d

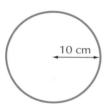
10 cm

_____

_____

_____

e

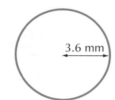
3.6 mm

_____

_____

_____

f

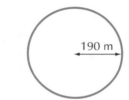
190 m

_____

_____

_____

2 Calculate the area of each of the following circles.
Leave all answers in terms of π.

a

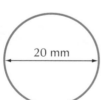
20 mm

_____

_____

_____

b

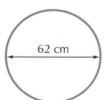
62 cm

_____

_____

_____

c

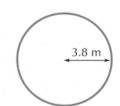
3.8 m

_____

_____

_____

Circles – area

3 Calculate the area of the following shapes.
Give your answers correct to two decimal places.

a

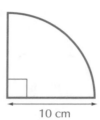

10 cm

b

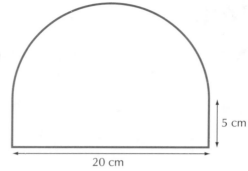

5 cm

20 cm

_____

_____

_____

_____

_____

_____

4 Calculate the shaded area in each of these shapes, correct to two decimal places.

a

10 cm

b

20 cm

c

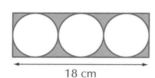

18 cm

_____

_____

_____

_____

_____

_____

_____

_____

_____

5 Work out the areas of the following shapes of semicircles.
Leave all answers in terms of π.

a

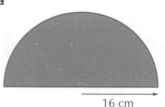

16 cm

b

8 cm

c

4 cm

_____

_____

_____

_____

_____

_____

d By looking at your answers to parts **a**, **b** and **c**, write down the area of eight semicircles each with a radii of 2 cm.

_____

Prisms – surface area

Calculate the surface area of each of the following prisms.
All lengths are in centimetres.

1

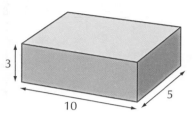

2

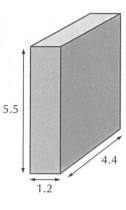

3

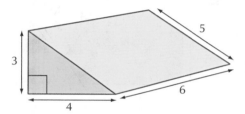

4

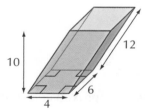

5

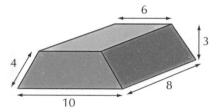

6

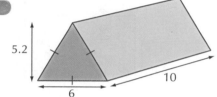

7

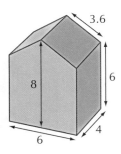

8

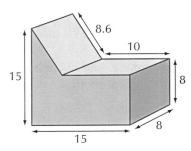

9

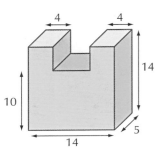

10

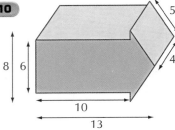

11 A4 paper is approximately 21 cm by 30 cm.
What is the maximum number of cylindrical paper tubes of diameter 3 cm and height 6 cm that can be made from a piece of A4 paper?

Prisms – volume

Calculate the volume of each of the following prisms.
All lengths are in centimetres.

1

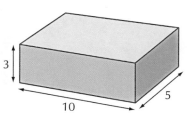

2

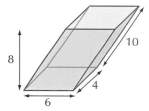

_____

_____

_____

_____

3

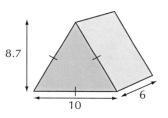

4

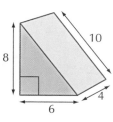

_____

_____

_____

_____

5

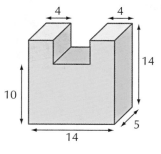

6

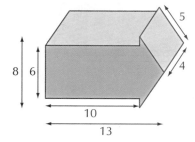

_____

_____

_____

_____

7

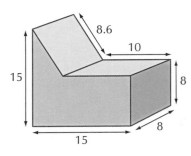

8

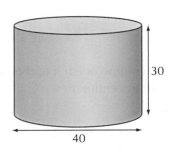

_____

_____

_____

9 A 25 m long hosepipe is made from plastic.
Its external diameter is 15 mm; its internal diameter is 12 mm.
Calculate the volume of plastic that makes the hosepipe.

Remember: Think of your units.

_____

_____

_____

_____

10 Which cake tin has the greatest volume? You must show your workings.

A

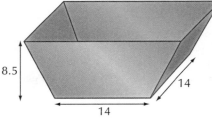

B

C

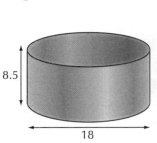

_____   _____   _____

_____   _____   _____

_____   _____   _____

_____   _____   _____

Answer: _____

Density ! Higher • Tier only

Remember:
Learn this!

M
D | V

1 Calculate the missing measurements in the table.

	Mass	Density	Volume
a	54 g	1.2 g/cm^3	
b		3.15 g/cm^3	60 cm^3
c	3.6 g		4.5 cm^3
d		0.85 g/cm^3	1 litre
e	13.38 kg	22.3 g/cm^3	
f	900 kg		1 m^3

2 The density of iron is 7900 kg/m^3. Work out the mass of 4.2 m^3 of iron.

3 Work out the volume of a liquid that has a mass of 0.6 kg and a density of 1.2 g/cm^3.

4 A 1 kg bag of sugar has a volume of 625 cm^3. Work out the density of the sugar in g/cm^3.

5 The density of aluminium kitchen foil is 2.4 g/cm^3. A roll of kitchen foil is 20 m long, 30 cm wide and 0.006 mm thick. Calculate the mass of the roll of foil.

6 A bar of pure gold has a density of 19.3 g/cm^3 and costs £24 825 per kilogram. The dimensions of a '400 oz' gold bar are 200 mm by 80 mm by 45 mm. How much is this gold bar worth?

Pythagoras

1 For each of the following triangles calculate the length, x.
Give answers correct to one decimal place where necessary.

a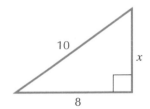

b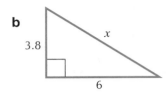

c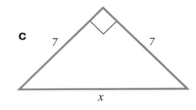

_____
_____
_____

2 For each of the following triangles calculate the length, x.
Give answers correct to one decimal place where necessary.

a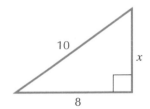

b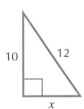

c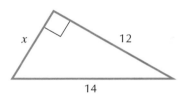

_____
_____
_____

3 Calculate the length of the diagonal of a square of side 10 cm.

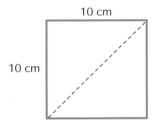

_____
_____
_____

4 Calculate the length of the diagonal of Stamford Bridge's football pitch.

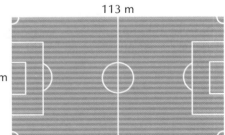

_____
_____

5 A ladder, 6 m long, leans against a wall.
The ladder reaches 5 m up the wall.
How far from the base of the wall is the foot of the ladder?

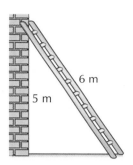

6 m
5 m

6 A right-angled triangle with shorter sides of 5 cm and 9 cm
is inside a circle with centre O.
Calculate the radius of the circle.

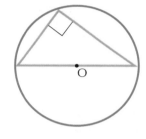

O

7 Calculate the length of the line segment AB, correct to the nearest whole number when:

a A is at point (3,4) and B is at (6,8)

b A is at point (−3,1) and B is at (4,−2).

8 A cuboid is cut through four of its vertices A, B, C and D leaving two identical pieces.
The diagram shows one of the pieces.

Calculate the distance AC, correct to two decimal places.

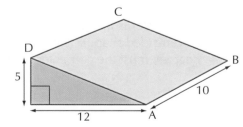

C
D
B
5
10
12
A

Regular polygons

1 For a regular pentagon calculate a) the interior angle and b) the exterior angle.

a _____ b _____

_____   _____

_____   _____

2 Explain why the internal angles of a hexagon equal 720°.

_____

_____

3 The diagram shows part of a regular polygon.
The external angle is 45°.
How many sides does the polygon have altogether?

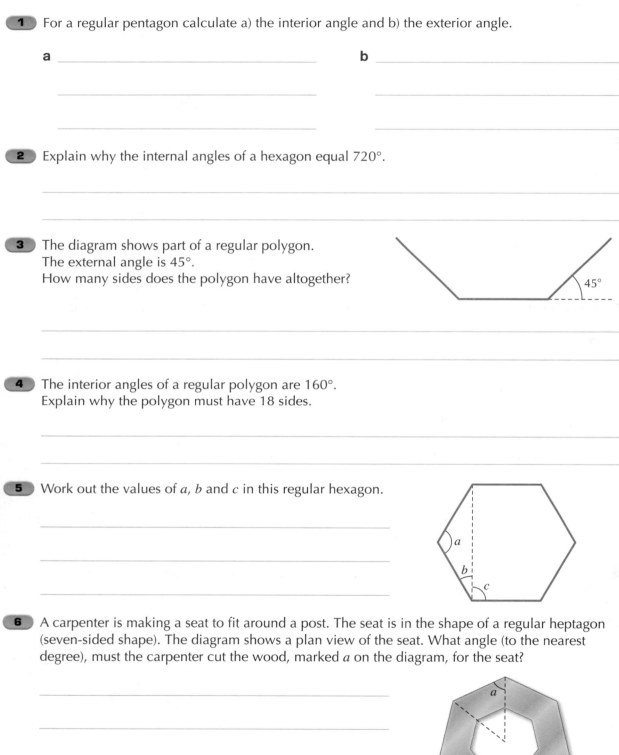

45°

_____

_____

4 The interior angles of a regular polygon are 160°.
Explain why the polygon must have 18 sides.

_____

_____

5 Work out the values of *a*, *b* and *c* in this regular hexagon.

_____

_____

_____

a

b

c

6 A carpenter is making a seat to fit around a post. The seat is in the shape of a regular heptagon (seven-sided shape). The diagram shows a plan view of the seat. What angle (to the nearest degree), must the carpenter cut the wood, marked *a* on the diagram, for the seat?

_____

_____

a

Translation

1

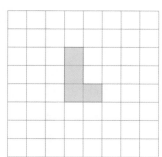 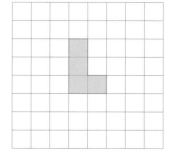

   a Translate the shape three squares right and one square down.

   b Translate the shape one square left and two squares up.

2 **a** Describe the translation that maps A onto C.

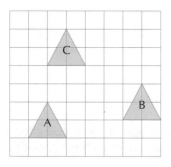

   b Describe the translation that maps B onto C.

   c Decribe the translation that maps C onto A.

3 Write down the column vector for each translation.

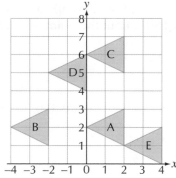

   a Shape B to shape A _____

   b Shape A to shape C _____

   c Shape D to shape A _____

   d Shape E to shape D _____

   e Shape C to shape B _____

   f Shape A to shape D _____

4 On a grid, $\begin{bmatrix} 3 \\ -1 \end{bmatrix}$ translates shape P onto shape Q and $\begin{bmatrix} 3 \\ 3 \end{bmatrix}$ translates shape Q onto shape R.

Write down the column vector that translates shape P directly onto shape R.

Rotation

1

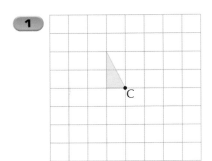

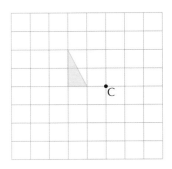

a Draw the image of the shape after a rotation of $\frac{1}{4}$ turn clockwise about centre C.

b Draw the image of the shape after a rotation of 180° about centre C.

2 **a** Draw the image of the shape after a rotation 90° clockwise about the point (0,0).

b Draw the image of the shape after a rotation 90° anticlockwise about the point (1, −3).

c Draw the image of the shape after a rotation 180° about the point (−1,−1).

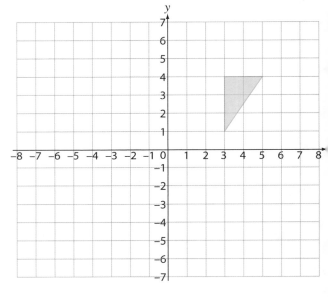

3 Rotate the shapes about centre (0,0) to make a pattern with rotational symmetry of order four.

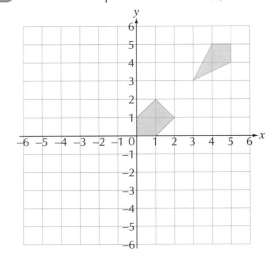

Reflection

1 Draw the reflection of each shape in the line $y = 1$.

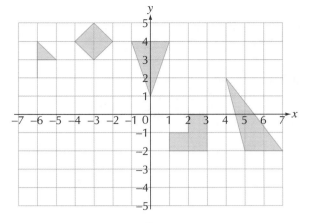

2 **a** Draw the line $y = x$ with a dashed line.

b Draw the reflection of shape A in the line $y = x$. Label the reflection shape W.

c Draw the reflection of shape B in the line $y = x$. Label the reflection shape X.

d Draw the reflection of shape C in the line $y = x$. Label the reflection shape Y.

e Draw the reflection of shape D in the line $y = x$. Label the reflection shape Z.

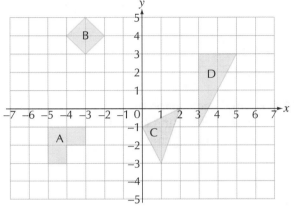

3 Describe the transformation that takes:

a Shape A to shape B.

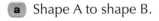

_____

_____

b Shape E to shape F.

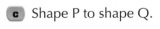

_____

_____

c Shape P to shape Q.

_____

_____

d Shape X to shape Y.

_____

_____

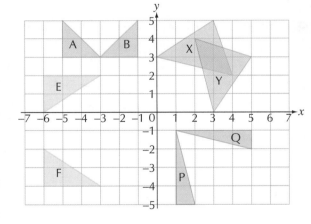

Enlargement

1 Enlarge the shape by a scale factor of three with (0,0) as the centre of enlargement.

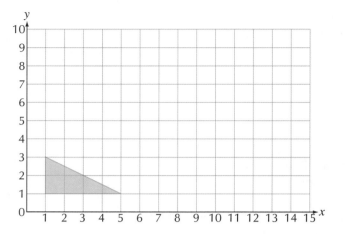

2 Enlarge each shape by a scale factor of two using *C* as the centre of enlargement.

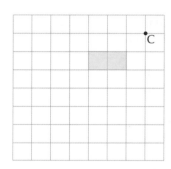

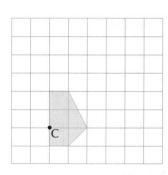

3 Enlarge each shape by the scale factor given, using *C* as the centre of enlargement.

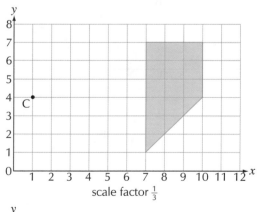

scale factor $\frac{1}{3}$

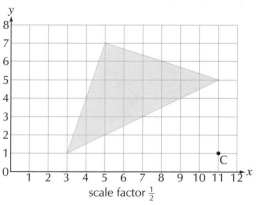

scale factor $\frac{1}{2}$

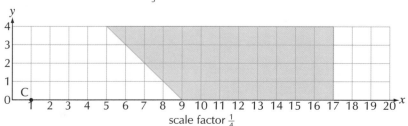

scale factor $\frac{1}{4}$

Constructions

1 Make an accurate drawing of each of these shapes.

a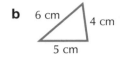

4 cm

50°

5 cm

b 6 cm   4 cm

5 cm

c

40° 55°

5 cm

d

5 cm   110°   5 cm

70°

5 cm

2 Using compasses and a ruler, construct an angle of 60° at point A.

A ●————————————

3 Using compasses and a straight edge, construct the perpendicular bisector of the line segment AB.

A ———————————————————————— B

4 Using compasses and a ruler, construct the angle bisector of ABC.

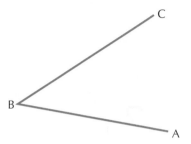

5 Using compasses and a straight edge, construct the perpendicular bisector of the line that passes through point A.

A

6 Using compasses and a ruler, construct a right angle from the line that passes through point A.

A ■

Loci

1 Draw the locus of points that are 2 cm from point X.

■ X

2 Draw the locus of points that are 2 cm from XY.

X —————————————— Y

3 Draw the locus of points that are 2 cm from this line.

Loci

4 A goat is tethered by a 4 m long rope to the corner of a shed. Show on the diagram of the shed all of the ground that the goat could reach. Let 1 cm = 1 m.

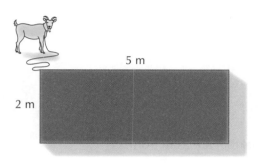

5 m

2 m

5

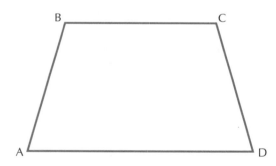

B　　　　　　　　　　C

A　　　　　　　　　　　D

a Construct the locus of points inside the trapezium that are equidistant from AD and CD.

b Construct the locus of points inside the trapezium that are 2 cm from AD.

c Construct the locus of points inside the trapezium that are 5 cm from A.

d Shade the region inside the trapezium where points are nearer to CD than to AD, within 2 cm of AD and more than 5 cm from A.

Dimensional analysis

1 A cuboid has sides of lengths a, b and c.

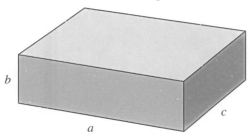

One of the formulae represents the total length of edges (L), one of them represents the total area of the faces (A) and one of them represents the total volume (V).

Which is which?

$2ab + 2ab + 2ab$ represents total _____

abc represents total _____

$4a + 4b + 4c$ represents total _____

2 For each of the following formulae indicate whether it represents a length (L), an area (A), a volume (V), is inconsistent (I) or is impossible (X).

The letters a, b, c, h, and r all represent a length. Two, three, four and π are all numbers and have no dimension.

a $\pi r^2 = $ _____

b $a + bc = $ _____

c $2a + 4b = $ _____

d $cab = $ _____

e $\dfrac{4}{3}\pi r^3 h = $ _____

f $a + b + c + h = $ _____

g $ab + h^2 = $ _____

h $\pi r = $ _____

i $a^2 b^2 = $ _____

j $\pi r^2 h = $ _____

k $a^2 b + b^2 c + rh = $ _____

l $4\pi^2 ar^2 = $ _____

3 State whether these formulae are a length (L), an area (A), a volume (V) or none of these (N).

a $\dfrac{a^2 + b^2}{r} = $ _____

b $a^3 - b^2 = $ _____

c $\pi r^2(r + 4h) = $ _____

d $3x(x + y) = $ _____

e $4x^2(y^2 - z) = $ _____

f $\dfrac{abc}{3r} = $ _____

Midpoints

1 Work out the midpoint for each of the line segments shown on the grid.

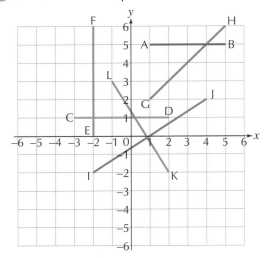

Midpoint of: AB = (_____, _____),   CD = (_____, _____),   EF = (_____, _____),

                  GH = (_____, _____),   IJ = (_____, _____),   KL = (_____, _____)

2 Work out the midpoints of the line segment AB when

a A = (0, 2) and B = (4, 0) _____

b A = (−2, 0) and B = (0, 4) _____

c A = (−2, −1) and B = (2, 5) _____

d A = (−3, −1) and B = (6, 6) _____

3 ABCD is a quadrilateral with coordinates A(2, −1), B(6, 4), C(2, 6), D(−2, 1).

a Work out the coordinates of the midpoint of the diagonal AC. _____

b Work out the coordinates of the midpoint of the diagonal BD. _____

4 The coordinates of the midpoint of a line segment AB are (2, 4).

Work out the coordinates of B when A is at

a (0, 0) _____

b (1, 2) _____

c (−1, 6) _____

Mean, median, mode and range

1 Here is a set of data: 7 8 3 6 7 8 5 2 7.

   a Calculate the range. _____

   b Write down the mode. _____

   c Find the median. _____

   d Calculate the mean. _____

_____

2 Here are the salaries for all eight people working in a small company.

£8 000,   £18 000,   £18 000,   £18 000,   £24 000,   £24 000,   £25 000,   £37 000

   a Calculate the range. _____

   b Write down the mode. _____

   c Find the median salary. _____

   d Calculate the mean salary. _____

_____

Everyone in the company gets a £2 000 pay rise. What is the new:

   e range in salaries? _____

   f modal salary? _____

   g median salary? _____

   h mean salary? _____

3 Look at the frequency table.

x	Frequency
2	7
3	3
4	4
5	7

   a Calculate the range. _____

   b Write down the mode. _____

   c Find the median. _____

   d Calculate the mean. _____

Mean, median, mode and range

4 Look at the frequency table.

Shoe size	3	4	5	6	7	8	9
Number of students	4	9	12	19	10	4	2

a Calculate the range. _____   **b** Write down the mode. _____

c Find the median. _____

d Calculate the mean. _____

_____

5 Look at the stem and leaf diagram.

3	0	4	7
4	8	8	9
5	4	6	7
6	2	4	

Key: 3 | 0 means 30

a Calculate the range. _____   **b** Write down the mode. _____

c Find the median. _____

d Calculate the mean. _____

_____

6 Look at the stem and leaf diagram.

1	7	7	7	
2	0	5	7	9
3	8	9		
4	0			

Key: 1 | 7 means 17 cm

a Calculate the range. _____   **b** Write down the mode. _____

c Find the median. _____

d Calculate the mean. _____

_____

7 Look at the grouped frequency table.

x	$0 < x \leq 5$	$5 < x \leq 10$	$10 < x \leq 15$	$15 < x \leq 20$
Frequency	16	27	19	13

a Write down the modal class. _____

b Calculate an estimate of the mean. _____

_____

8 Look at the grouped frequency table.

x	Frequency
$0 < x \le 10$	4
$10 < x \le 20$	9
$20 < x \le 30$	17
$30 < x \le 40$	13
$40 < x \le 50$	7

a Write down the modal class. _____

b Calculate an estimate of the mean. _____

_____

c Why can't we work out the range for this data?

_____

_____

9 A list of nine numbers has a mean of 4.8
What number must be added to give a new mean of five?

_____

_____

10 The mean weight of five basketball players is 68.4 kg. The mean weight of the five basketball players and their coach is 71 kg.
What is the weight of the coach?

_____

_____

11 The mean weight of a team of 15 rugby players is 78.5 kg. A player who weighs 71.2 kg leaves the team and is replaced by a player who weighs 75.7 kg.
What is the new mean weight of the team?

_____

_____

Mean, median, mode and range

12 Students at a school are awarded 'merits' for good work. This table shows the number of merits awarded to the students in three different year groups in one month.

Number of merits	Year 7 frequency	Year 8 frequency	Year 9 frequency
0	14	4	24
1	17	22	38
2	42	30	20
3	24	33	13
4	8	17	11
5	4	16	7
6	15	8	10

a How many students are in year 7? _____

b Calculate the mean number of merits awarded per student for the students in year 7. Give your answer correct to two decimal places.

_____

_____

c The year group with the highest mean number of merits awarded per student is awarded a prize. Which year group won the prize?

_____

_____

13 The results of a survey about the number of homeworks given to some students are shown in a frequency table. The frequency in the last row has been covered by a tea stain!

The median number of homeworks is 2.5
Calculate the mean number of homeworks given.
Give your answer correct to two decimal places.

Homeworks given	Frequency
0	6
1	3
2	8
3	7
4	

_____

_____

_____

_____

_____

Correlation

1 Miss Directed is investigating this claim about Ford motor cars:

'The older the car, the less it is worth.'

Miss Directed collects some data from a local newspaper and draws a scatter diagram to show her results.

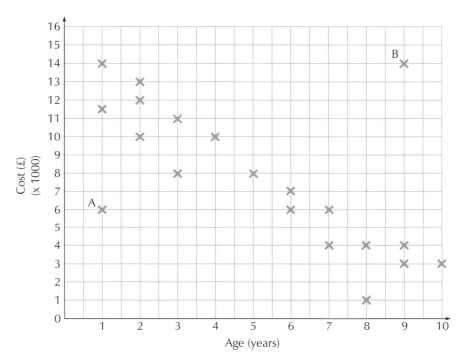

Miss Directed sees three more cars to add to her data: three years old, valued at £10 000

three years old, valued at £9 000

seven years old, valued at £5 500

a Plot these points onto the scatter diagram.

b Decide whether the claim 'the older the car, the less it is worth', is correct.

_____

c Cars A and B do not fit the general trend. What can you say about

i car A? _____

ii car B? _____

d Do you think that this general trend continues as cars get even older? Explain your answer.

_____

Correlation

2 Mr Stephens is investigating this claim:

'The greater the percentage attendance at his maths
lessons, the higher the mark in the end of year maths test.'

Mr Stephens collects data from his class and draws a scatter diagram to show his results.

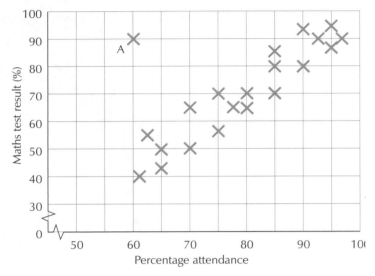

a Decide whether the claim 'the greater the percentage attendance at his maths lessons, the higher the mark in the end of year maths test', is correct or not.

_____

b Student A does not fit the general trend. What can you say about student A?

_____

c Describe the correlation between percentage attendance and the maths test result.

_____

d Draw a line of best fit on the scatter diagram.

e Use your line of best fit to estimate

   i   the maths result of a student with an 85% attendance   _____

   ii the attendance of a student with a test result of 60%.   _____

f Why would it not be useful to use the line of best fit to estimate the test mark for a student with a 20% attendance?

_____

3 The table shows the score of some pupils in a mental test and a written test.

Pupil	Amy	Billy	Chris	Dom	Eddy	Flynn	Glyn	Harjit	Iris	Jill
Mental	13	17	22	25	20	14	14	18	12	21
Written	25	30	34	13	37	29	25	44	19	35

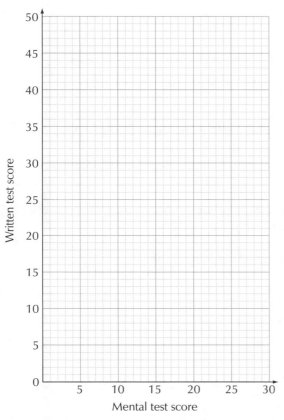

a Plot the data on a scatter graph using the axes above.

b Draw a line of best fit.

c Keith missed the mental test, but scored 32 on the written test. Estimate the score that Keith would have got on the mental test.

_____

d One person did not do as well as expected on the written test. Who do you think that was? Give a reason for your choice.

_____

e Describe the correlation between the two types of test.

_____

Retail price index

1 In 2004, the price paid to farmers for a litre of milk was 18.5p. Using 2004 as a base year, the price index of milk in each of the next five years is shown in the table.

Year	2004	2005	2006	2007	2008	2009
Index	100	101	97	109	134	124
Price	18.5					

Work out the amount paid to farmers for a litre of milk in each subsequent year.
Give your answer correct to one decimal place.

2 The 1975 cost of an ounce of gold was $160. Taking 1975 as the base year of 100, work out how much one ounce of gold cost in January 2010, when the price index was 715.

3 A paper plate factory produced 341 600 packs of paper plates in May 2010. This represented an index of 61 using May 2000 as the base year with an index of 100.

How many packs of paper plates did the factory produce in May 2000?

4 The graph shows the exchange rate for the dollar against the pound in 2009.

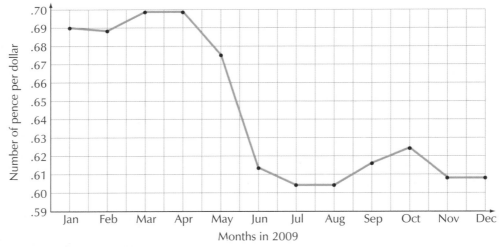

Months in 2009

a What was the exchange rate in January 2009?

b Using January 2009 as the base of 100, work out the approximate index number for December 2009.

Questionnaires

1 These are questions from a questionnaire on healthy living.

> Eating lots of vegetables everyday is good for you. Don't you agree?
>
> Strongly agree ☐   Agree ☐   Not sure ☐

Give two reasons why this question is unsuitable.

_____

_____

> Do you eat fast food?   Yes   No
>
> If 'yes', how many times a week, on average, do you eat fast food?
>
> Once or less ☐   2 or 3 times ☐   4 or 5 times ☐   more than 5 ☐

Give two reasons why this is a good question.

_____

_____

2 Andrew is carrying out a survey about how much pocket money students in his school get each week. One of his questions is 'How much pocket money, on average, are you given each week?'

a Design a response section for Andrew's question.

b Write a question that he can use to find out what students spend their pocket money on.

_____

_____

Andrew decides to use one of three methods to do the survey.
Method 1: Ask the first 50 students he sees on Monday break time.
Method 2: Choose 50 students at random.
Method 3: Choose 26 students, picking one whose surname begins with each letter of the alphabet.

c Give a reason why method 3 is not suitable.

_____

d Which of the other two methods for doing the survey will give the most reliable results? Give a reason for your choice.

_____

Probability

1 Nemah throws a dice and records the number of threes that he gets after various numbers of throws.

Number of throws	10	50	100	200	500	1000
Number of threes	3	6	19	30	92	163
Experimental probability						

    a Complete the table by calculating the experimental probability of a three at each stage that Nemah recorded his results.

    _____

    _____

    _____

    b What is the theoretical probability of throwing a three?

    _____

    c If Nemah threw the dice 6000 times, how many threes would you expect him to get?

    _____

2 A coin is flipped 1000 times. How many times should 'heads' land face up?

_____

3 A bag contains 30 balls, 15 of which are blue, 10 are red and five are green. A ball is taken out at random and then replaced. This is repeated 300 times. How many times would I expect to get:

    a A red ball? _____

    b A blue or a green ball? _____

    c A ball that is not green? _____

    d A yellow ball? _____

4 Crawford and Janet each carry out an experiment with the same dice. The tables show their results.

Crawford's results

Number	1	2	3	4	5	6
Frequency	6	4	13	9	10	8

Janet's results

Number	1	2	3	4	5	6
Frequency	53	48	45	54	50	52

Crawford thinks the dice is biased. Janet thinks the dice is fair. Who is correct? Explain your choice.

_____

_____

5 A card is taken at random from a shuffled pack of cards.

a What is the probability it is red? _____

b What is the probability it is black? _____

c Explain why the events in part **a** and **b** are mutually exclusive.

_____

_____

d Explain why the events in part **a** and **b** are exhaustive.

_____

_____

6 **a** List the possible outcomes when three coins are thrown together. One has been done for you.

HHH, _____

b What is the probability of getting three heads? _____

c What is the probability of getting two heads and one tail? _____

7 A fair six-sided dice is rolled twice.
What is the probability of getting a one and then a two?

_____

Probability

8 Two four-sided dice are rolled together and their scores added.

    a Draw a sample space diagram showing the possible outcomes.

    b What is the probability of getting a total of six? _____

9 When a coin is flipped and an ordinary dice is rolled together, what is the probability of getting a tail on the coin and a two on the dice?

_____

_____

10 The probabilities of whether students, picked at random, are vegetarian or not are shown in the table.

	Boys	Girls
Vegetarian	0.07	0.2
Not vegetarian	0.42	0.31

    a What is the probability that a student chosen at random is a vegetarian? _____

    b There are 170 girls in the school who are vegetarian. How many students are there in the school altogether?

_____

    c How many non-vegetarian boys are in the school altogether?

_____

Revision – Paper 1

1 Use approximations to estimate the value of the following.

a $\dfrac{1.8 \times 51.3}{7.3 - 2.21}$

b $\dfrac{415.1 + 47.49}{5.383 \times 1.922}$

_____

_____

_____(3)

_____

_____

_____(3)

2 Here is a pattern of lines.

pattern 1   **pattern 2**   **pattern 3**

Work out how many lines will be needed for the nth pattern. (2)

_____

3 The nth term of a sequence is $2n + 8$.

a Show that all the terms in the sequence are even. (2)

_____

_____

b Which term of the sequence is 26? (2)

_____

c Find the nth term of this fractional sequence.

$\dfrac{1}{35}, \dfrac{5}{32}, \dfrac{9}{29}, \dfrac{13}{26}, \dfrac{17}{23}, \cdots$ (4)

_____

_____

4 Josie says 'All the terms in the sequence with *n*th term $n^2 + 2$ are even.'
Give a counter-example to show that she is wrong. (1)

_____

5 A botanist measures the heights of some young plants.
He writes that the heights of the plants, *h*, lie in the interval:

$$30\,cm \leq h < 35\,cm$$

a Write down all the whole number heights that the plants could be. (1)

_____

b Show the inequality $30 \leq w < 35$ on this number line.

27   28   29   30   31   32   33   34   35   36   37
(1)

6 Solve $4a - 3 < 17$. (2)

_____

_____

7 **a** Write 540 as a product of its prime factors in index form. (3)

_____

_____

_____

_____

_____

_____

b Clive is tiling his kitchen floor. The kitchen is a rectangle measuring 540 cm by 315 cm. He wants to use identical square tiles to completely cover the floor with no overlap. What is the largest size of square tile that he can use? (2)

_____

_____

_____

_____

8 **a** Work out $\sqrt[3]{12^2 - 19}$. (1)

_____

_____

b Carys writes $6^{15} \div 6^5 = 6^3$
Is Carys correct? Give a reason for your answer. (1)

_____

_____

9 In 2008 a florist sold 1000 orchids for £10 each.
In 2009 the florist sold 20% more orchids than he did in 2008, but each orchid sold for 20% less than they did in 2008.
Will the florist take the same amount of money in 2009 as in 2008?
You **must** show your working. (4)

_____

_____

_____

_____

_____

10 Osama is tiling his kitchen. The kitchen is a rectangle measuring $3\frac{1}{2}$m by $5\frac{3}{4}$m.
Tiles come in packs that cost £15 per pack. Each pack covers an area of $1\,\text{m}^2$.
Just in case he breaks some tiles, Osama decides to buy more packs than he needs. He buys an extra 10% of the tiles.

How much does Osama pay for his tiles? (5)

_____

_____

_____

_____

_____

_____

11 **a** Complete the table of values for $y = x^2 - 2x + 2$ for values of x from -3 to $+3$. (2)

x	-3	-2	-1	0	1	2	3
y	17		5	2		2	

b Draw the graph of $y = x^2 - 2x + 2$ for values of x from -3 to $+3$. (3)

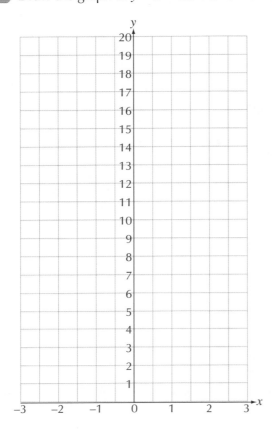

c Use your graph to estimate the value of y when $x = 2.5$. (1)

_____

d Use your graph to solve the equation $x^2 - 2x + 2 = 4$. (2)

_____

12 **a** A ship leaves port, P, and sails on a bearing of 220° for 6 km to a point A. At point A the ship changes direction and sails 8 km on a bearing of 075° to a point B.

   i Using a scale of 1 cm = 1 km, draw a diagram to show the ship's journey. (2)

   ii Measure and write down the distance and the bearing the ship must sail on to return directly from point B to port. (2)

A •

b Using only compasses and a ruler, construct the perpendicular from the point B to the line. You must show your construction lines.

• B

(2)

13 Triangle A is drawn on the grid.

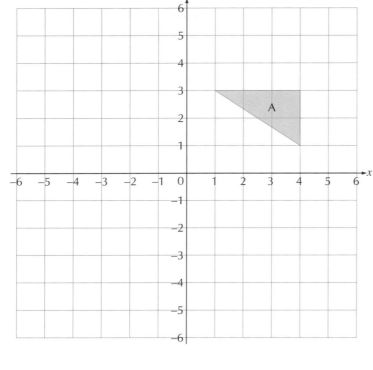

a Draw the image of triangle A after a rotation of 90° anticlockwise about the point (0, 1). Label the image B. (3)

b Triangle A is reflected to form a new triangle C. The vertices of triangle C are at (4, −3), (4, −5) and (1, −5). Work out the equation of the mirror line. (2)

_____

_____

c Enlarge triangle A by scale factor $\frac{1}{2}$, using (−4, −3) as the centre of enlargement. Label the image D. (2)

d Draw the image of triangle A after a translation by the vector $\begin{bmatrix} -4 \\ -6 \end{bmatrix}$. Label the image E. (2)

Revision – Paper 2

1 An estate agent collects information on the average prices of one-bedroom flats at certain distances from the local train station. The table shows her results.

Distance from train station (miles)	1	7	3	9	16	12
Average price (£000s)	162	139	156	130	118	124

a On the graph paper below, draw a scatter diagram to show this information. (2)

b Write down the type of correlation shown by your scatter diagram. (1)

_____

c Draw a line of best fit on your scatter diagram. (1)

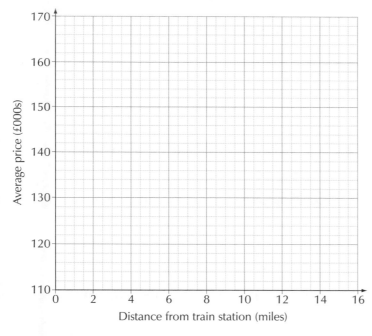

d Jordan needs to catch the train to go to work. She has £150 000 to spend on a one-bedroom flat. Use your line of best fit to estimate how close to the train station she could buy a flat. (1)

_____

e Why would it not be useful to use the line of best fit to estimate the cost of a one-bedroom flat 24 miles from the railway station? (1)

_____

2 A supermarket recorded the number of minutes 102 shoppers spent in the chocolate aisle before choosing an Easter egg. The table shows the results.

Number of minutes, m	Frequency
$0 < m \leq 1$	20
$1 < m \leq 2$	32
$2 < m \leq 3$	16
$3 < m \leq 4$	12
$4 < m \leq 5$	20

a Calculate an estimate of the mean shopping time. (4)

b Explain why you can't work out an exact mean shopping time. (1)

c Work out the class interval containing the median. (2)

d The supermarket manager writes a report on the shopping times in the chocolate aisle. He writes:

'Over 50% of shoppers chose their Easter egg within 2 minutes.' (1)

Is the supermarket manager correct? Give a reason for your answer.

3 Tim wants to find out information about the videogames his classmates play. He has written this question for his survey.

> How long do you play videogames in one week? Please tick one box only.
>
> 2 to 4 hours ☐   4 to 6 hours ☐   6 to 8 hours ☐

a Write down two criticisms of the question. (2)

1: _____

2: _____

b Re-write the question to make it more suitable. (2)

_____

_____

Tim has also written this question for his survey.

> The best games consul is X-box. Don't you agree?
>
> Strongly agree ☐   Agree ☐   Don't know ☐

c Give 2 criticisms of the question. (2)

1: _____

2: _____

d Rewrite the question to make it more suitable. (2)

_____

_____

4 **a** The amounts spent by 50 customers at a shop one Saturday in October are shown.

Amount spent, x (£)	Frequency
$0 \leq x < 10$	23
$10 \leq x < 20$	14
$20 \leq x < 30$	9
$30 \leq x < 40$	4

a Work out the probability that a customer chosen at random spent more than £20. (2)

_____

b Explain why it is not possible to work out the probability that a customer chosen at random spent exactly £15. (1)

_____

c Altogether the shop had 2000 customers in October. How many of them would you expect to spend between £20 and £30? (2)

_____

5 The cost of a new lorry is £22 000 + $17\frac{1}{2}$% VAT. Pat decides to pay by credit. She pays an initial deposit of 15% and then 24 monthly payments of £1000. How much more does Pat pay by buying on credit? You must show your working. (4)

6 Farquar buys an old painting for £850. He spends £125 on having it cleaned then sells the painting for £1560. Farquar says that he makes a 50% profit. Is Farquar correct? You must show your working. (3)

7 A biologist estimated that a termite mound had 520 000 termites in 2009. She also estimated that the mound had 546 000 termites in 2010. She says:

'If the number of termites continues to increase by the same percentage each year, by 2015 there will be around 700 000 termites in the mound.'

Do you agree? You must show working to support your answer. (3)

8 Farrouq and Mahmoud are brothers. Each year they share a £1000 gift from their uncle in the ratio of their ages. This year Farrouq is 6 years old and Mahmoud is 2 years old. Show that in four years time the difference between the amounts they receive will be halved. (3)

9 A gas supplier uses this formula to work out the total cost of the gas a customer uses.

C = 0.04G + 30 C is the total cost of the gas in pounds.
G is the number of units of gas used.

a Neal uses 600 units of gas. What is the total cost of the gas he uses? (2)

b The total cost of the gas Sian uses is £74. How many units of gas does she use? (2)

10 **a** Solve the equation $7x + 12 = 39 - 3x$ (3)

b Expand $x(3x - 4)$ (2)

c Expand and simplify $4(3d + 2) - 3(d - 2)$ (3)

11 The diagram shows a hollow triangular prism with no end faces. The surface area of the prism is $(9x + 3)\,cm^2$.

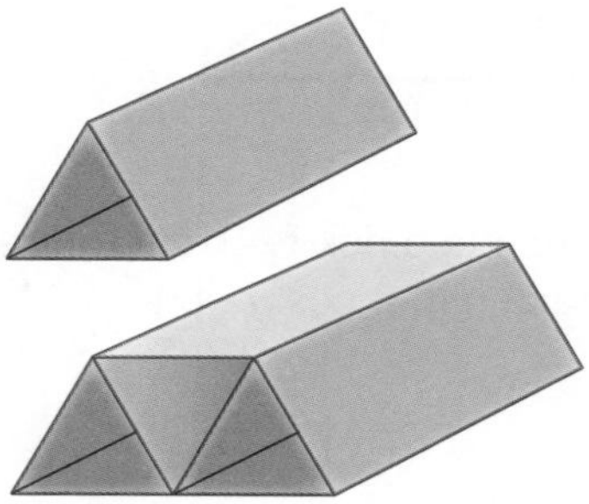

Three of these prisms are joined to make the trapezium shown. The surface area of the trapezium is $50\,cm^2$. Work out the value of x. (5)

12 The diagram shows a regular octagon.

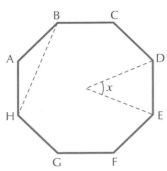

a Work out the size of the angle marked x. (2)

_____

b Work out the size of the interior angle of a regular octagon. (2)

_____

c Work out the size of angle ABH. (2)

_____

13 Bob is going to make a raised vegetable plot by putting concrete blocks around a triangular piece of land in the corner of his garden. The diagram shows the dimensions of the triangle of land.

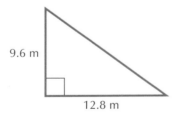

Each concrete block is 40 cm long and costs £1.50 to buy.
How much will it cost Bob to put the concrete blocks around the triangular piece of land? (6)

_____

_____

_____

_____

14 This triangular prism has a volume of 54 cm³.
Work out the length, x, of the prism.
You must show your working.

5 cm   (5)

x cm

3 cm

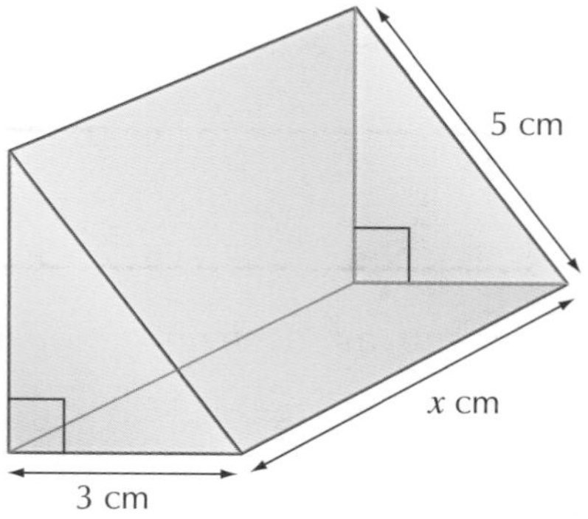

15 a A semicircle of radius 10 cm and a square have the same perimeter.
What is the side length of the square? Give your answer correct to two decimal places.
You must show your working.   (3)

b A circle of diameter 30 cm and a square have the same perimeter.
Which one of them has the greater area?
You must show your working.   (5)

16 The area of this rectangle is 160 cm^2.

$x^2 + x$ cm

x cm

Use trial and improvement to find the value of x correct to 1 decimal point.   (5)

Answer: _____

17

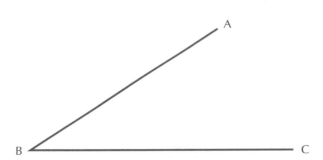

Find and shade the region of points that satisfy all three of the following conditions.
i   The points are nearer to AB than BC.
ii The points are nearer to B than C.
iii The points are not further than 5 cm from C.   (4)

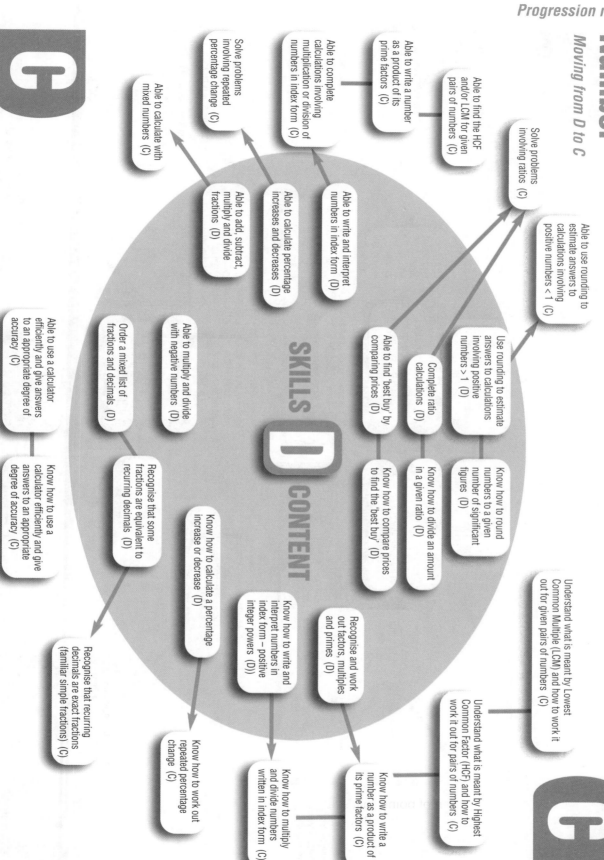

Number
Moving from D to C

C

SKILLS D CONTENT

Able to write a number as a product of its prime factors (C)

Able to find the HCF and/or LCM for given pairs of numbers (C)

Able to complete calculations involving multiplication or division of numbers in index form (C)

Solve problems involving repeated percentage change (C)

Able to calculate with mixed numbers (C)

Able to calculate percentage increases and decreases (D)

Able to write and interpret numbers in index form (D)

Able to add, subtract, multiply and divide fractions (D)

Solve problems involving ratios (C)

Able to use rounding to estimate answers to calculations involving positive numbers < 1 (C)

Use rounding to estimate answers to calculations involving positive numbers > 1 (D)

Complete ratio calculations (D)

Able to find 'best buy' by comparing prices (D)

Able to multiply and divide with negative numbers (D)

Order a mixed list of fractions and decimals (D)

Recognise that some fractions are equivalent to recurring decimals (D)

Know how to round numbers to a given number of significant figures (D)

Know how to divide an amount in a given ratio (D)

Know how to compare prices to find the 'best buy' (D)

Know how to calculate a percentage increase or decrease (D)

Know how to write and interpret numbers in index form – positive integer powers (D))

Recognise and work out factors, multiples and primes (D)

Able to use a calculator efficiently and give answers to an appropriate degree of accuracy (C)

Know how to use a calculator efficiently and give answers to an appropriate degree of accuracy (C)

Recognise that recurring decimals are exact fractions (familiar simple fractions) (C)

Know how to work out repeated percentage change (C)

Know how to multiply and divide numbers written in index form (C)

Know how to write a number as a product of its prime factors (C)

Understand what is meant by Lowest Common Multiple (LCM) and how to work it out for given pairs of numbers (C)

Understand what is meant by Highest Common Factor (HCF) and how to work it out for pairs of numbers (C)

PROGRESSION MAP

Algebra
Moving from D to C

C

SKILLS D CONTENT D

Know how to simplify linear inequalities to find a solution set (C)

Know how to convert a problem given in words into a linear equation or formula (C)

Know that formulae can be rearranged using arithmetic conventions (C)

Know how to expand and simplify algebraic expressions using BODMAS/BIDMAS (C)

Know notation for inequalities and that they represent a range of values (C)

Know how to solve an equation with the variable in a bracket (D)

Know how to identify common factors and factorise linear expressions (D)

Know how to substitute numbers into expressions and evaluate using arithmetic conventions (D)

Know how to plot a quadratic graph by substituting values (C)

Know how to find approximate solutions to equations by using a quadratic graph (C)

Know how to find approximate solutions to equations by trial and improvement, and when it's appropriate to use this method (C)

Know how to tackle linear equations where the variable appears on both sides of the equals sign (D)

Know how to solve a linear equation where the variable occurs in the numerator of a fraction (D)

Know how to expand a linear bracket (D)

Know that straight line graphs can be written algebraically in terms of x and y (D)

Know that the rule for generating a sequence can be expressed in terms of its nth term (D)

Know how to find an algebraic rule for the nth term of a linear sequence (C)

Use trial and improvement to solve equations when appropriate (C)

Able to solve linear equations where the variable appears on both sides of the equals sign (D)

Can solve linear equations where the variable occurs in the numerator of a fraction (D)

Able to expand linear brackets accurately (D)

Able to draw straight line graphs from equations by plotting points (D)

Able to substitute numbers into an nth term rule (D)

Can find the algebraic rule for the nth term of a linear sequence (C)

Able to rearrange simple formulae to solve problems (C)

Able to solve simple linear equations that include the variable in a bracket (D)

Able to factorise simple linear expressions (D)

Able to construct a simple formula from a rule given in words (C)

Can construct and solve a linear equation from problem given in words (C)

Able to solve simple linear inequalities and represent solutions on a number line (C)

Able to expand and simplify algebraic expressions (C)

Able to draw quadratic graphs using a table of values and use graphs to solve problems (C)

C

Geometry and Measures

Moving from D to C

SKILLS D CONTENT

Able to reflect a 2D shape in the line $y = x$ or $y = -x$ (C)

Able to rotate a 2D shape about any point (C)

Enlarge a 2D shape from any centre (C)

Enlarge a 2D shape by a fraction scale factor (C)

Able to translate a 2D shape by a vector (C)

Work out unknown sides of shapes using scale factors and ratios (C)

Enlarge a 2D shape by a whole number scale factor (D)

Calculate the area of a trapezium (D)

Reflect a 2D shape in a line $x = a$ or $y = b$ (D)

Rotate a 2D shape about the origin (D)

Solve problems in 2D using Pythagoras' theorem (C)

Use Pythagoras' theorem in right-angled triangles (C)

Understand how Pythagoras' theorem works in right-angled triangles (C)

Calculate the volume of prisms and cylinders (C)

Able to construct line and angle bisectors (C)

Find interior and exterior angles in polygons (C)

Draw and describe the locus of a point from a given rule (C)

Solve problems using loci (C)

Solve problems involving the perimeter, area or volume of simple shapes (D)

Calculate the area and circumference of a circle (D)

Work out simple compound measures (speed) (D)

Draw plans and elevations of 3D shapes (D)

Know how to combine measures into simple compound measures (e.g. speed) (D)

Know how to read and use map scales (D)

Understand how a point, line or region can be identified by rules (loci) (C)

Know the formulae for the area and circumference of a circle (D)

Know how to work out the perimeter, area or volume of simple compound shapes (D)

Know properties and terminology of angles in parallel lines (D)

Use the formula for the area of a trapezium (D)

Find angles in triangles and quadrilaterals (D)

Understand the effect of an enlargement on perimeter, area or volume (C)

Understand and show simple geometric proofs (C)

Identify interior and exterior angles in polygons (C)

Know why two shapes are similar and how similarity can be used to solve problems (C)

Know transformations of 2D shapes and how they can be described (C)

Understand and use vector notation to describe translations (C)

C

Statistics and Probability

Moving from D to C

SKILLS D CONTENT

Design questionnaires and surveys (C)

Understand how trends can be used to identify unlikely or extreme results (C)

Know how to select group sizes for data and describe groups clearly using inequality symbols (C)

Identify bias in a questionnaire question (D)

Know how to design a questionnaire question with response choices (D)

Understand correlation and how links between data can be identified from a scatter diagram (D)

Know how to use probability to calculate relative frequency (C)

Understand the loss of accuracy in some calculations when data has been grouped (C)

Know when and how to group data in a frequency table (D)

Understand and distinguish between primary and secondary data (D)

Calculate the probability of an event happening when you know the probability that the event doesn't happen and that the total probability of all possible outcomes is 1 (D)

Understand the difference between experimental and theoretical probability (C)

Find an estimate of the mean from a grouped frequency table of continuous data (C)

Find the mean from a frequency table of discrete data (D)

Draw an ordered stem-and-leaf diagram (D)

Identify data needed to solve a problem (D)

Draw a frequency polygon for continuous data (C)

Draw a frequency polygon for discrete data (D)

Solve problems involving two way tables (D)

Draw a line of best fit on a scatter diagram (D)

Use mean and range to compare two sets of data (D)

Able to predict the expected number of successes from a given number of trials if you know the probability of one success (D)

Calculate the relative frequency from experimental evidence and compare this with theoretical probability (C)

Able to interpret a line of best fit (C)

Recognise and describe the different types of correlation (C)

Express probabilities using fractions, decimals or percentages (C)

C

ANSWERS TO NUMBER

Rounding / Giving an approximate answer

1 a 6   **b** 0.06   **c** 800

2 a 22   **b** 0.87   **c** 57000

3 a 91.0   **b** 0.0379   **c** 7900

4 a 4000   **b** 7.4   **c** 0.00556

   d 99.9   **e** 100   **f** 100

5 a 72 800   **b** 36.0   **c** 1880

6 a $70 \times 20 = 1400$ **b** $200 \div 40 = 5$

   c $\frac{30 \times 10}{0.5} = 600$ **d** $\frac{9 \times 5}{15} = 3$

   e $\sqrt{81} = 9$ **f** $\frac{4000}{0.5} = 80\,000$

   g $\frac{600}{15} = 40$ **h** $\frac{600}{0.4} + \frac{200}{0.5} = 1900$

   i $\frac{6 \times 20}{20} = 6$ **j** $0.2 - 0.6 - 0.9 = -1.3$

Prime factors / LCM / HCF

1 a $2^2 \times 7$ **b** $2^3 \times 3^2$

2 a 35   **b** 504

3 a 4   **b** 4

4 40 minutes

Indices

1 a 3^5 **b** 10^3 **c** 2^6

2 a a^4 **b** b^2 **c** x^6

3 a 5^4 **b** 7^8 **c** 2^{20} **d** 3^6

   e 6 (or 6^0)   **f** 8^5 **g** $y4$ **h** b^8

   i j^{20} **j** m^4 **k** a **l** 1 (not t^0)

   m 7^2 **n** x^3 **o** x^{-3} **p** d^9

   q $4a^9$ **r** h^6 **s** $3x^4$ **t** $3b$

   u $2t^3$

Fractions

1 a $\frac{25}{28}$ **b** $\frac{38}{45}$ **c** $4\frac{14}{15}$ **d** $1\frac{19}{40}$

   e $\frac{7}{30}$ **f** $\frac{5}{36}$ **g** $5\frac{11}{12}$ **h** $4\frac{5}{12}$

   i $6\frac{19}{40}$ **j** $5\frac{3}{8}$ **k** $\frac{3}{8}$ **l** $\frac{2}{3}$

   m $4\frac{3}{8}$ **n** 8   **o** $10\frac{1}{2}$ **p** $33\frac{3}{5}$

   q $\frac{3}{4}$ **r** $1\frac{1}{3}$ **s** $10\frac{6}{25}$ **t** $\frac{1}{4}$

2 $7\frac{3}{20}$ m²   **3** 15 strides

Percentage increase an ase

1 £280   **2** £2.80   **3** 39.6 m   **4** 3.96 m

5 4%   **6** £14 687.50

7 a Decreased **b** -3.5 cm²   **8** £371.30

One quantity as a percentage of another

1 a 20%   **b** 25%   **c** 53.3%

   d 10%   **e** 37.5%   **f** 8.3%

2 English = 60%, History = 62.2% so Jodie is better at History than English.

3 20%   **4** 9.4%   **5** 78.9%   **6** 76.1%

Compound interest

1 £73.50   **2** £1.50

3 $2000, invested for 3 years, at a rate of 5% per year.

4 £1192.52   **5** Approx 540 000

6 a 4.07 m   **b** 13 years

7 £10 555   **8** Approx 600 000

Sharing an amount by a given ratio

1 20:30   **2** 160:240   **3** £27:£63

4 10 m:990 m   **5** 20 ml:30 ml:50 ml

6 6.75 kg:5.625 kg:1.125 kg

7 500 kg cement, 1000 kg sand, 2000 kg gravel

8 a Aaron £1188, Beth £1782, Charlie £2970

   b Aaron, by £72

9 40°

Ratio

1 4 male teachers   **2** 60 teachers

3 38 staff

4 a 315   **b** 630

5 20 litres lemonade, 0.5 litres ginger

6 £69.75

Basic powers and roots

1 16, 25, 36, 49, 64, 81, 100, 121, 144, 169, 196, 225

2 27, 64, 125

3 a 10   **b** 12   **c** 4

   d 1   **e** 10   **f** 12   **g** 3

4 4 and -4

5 $-10 \times -10 = +100$

6 a 26   **b** 302.76   **c** 3.1   **d** 0.343

   e 0.008   **f** 16

7 a 2   **b** 6

8 a 29   **b** 117   **c** 7

Reciprocals

1 a i $\frac{1}{2}$ **ii** 0.5   **b i** $\frac{1}{5}$ **ii** 0.2

   c i $\frac{1}{10}$ **ii** 0.1   **d i** $\frac{1}{100}$ **ii** 0.01

2 Decimal becomes smaller.

3 a 4   **b** 0.4   **c** 0.04   **d** 40

4 a $1\frac{2}{7}$ **b** $\frac{9}{16}$ **c** $\frac{9}{25}$ **d** $\frac{9}{34}$

5 a 2.5   **b** 0.4   **c** 1   **d** 1   **e** yes

6 For example, the reciprocal of 0.1 is 10.

Answers

Limits

1 a 4.5 cm ≤ 5 cm < 5.5 cm   **b** 45 km/h ≤ 50 km/h < 55 km/h

c 45 students ≤ 50 students ≤ 54 students

d £45 ≤ £50 ≤ £54.99

e 295 m/s ≤ 300 m/s < 305 m/s

f 250 puppies ≤ 300 puppies ≤ 349 puppies

2 a 24.5 cm ≤ 25 cm < 25.5 cm

b 22.5 cm ≤ 25 cm < 27.5 cm

c 47 495 fans ≤ 47 500 fans ≤ 47 504 fans

d 12.65 cm ≤ 12.7 cm < 12.75 cm

3 a 495 and 504   **b** 495 g and 505 g

c Mass is a continuous measurement, but sweets are counted individually.

4 a 40.5 cm (40.499 999 ...cm)

b 79.5 cm   **c** 3260.25 cm²

ANSWERS TO ALGEBRA

Factorising

1 a $6(a + 2)$ **b** $4(a + 2b)$ **c** $2(2x + 3y)$

d $2(4t - 3p)$ **e** $2a(b + 3c)$ **f** $5m(n - p)$

g $p(p + 5)$ **h** $h(7 - h)$ **i** $x(3x + 2)$

j $3t(t - p)$ **k** $3x(2x + 3y)$ **l** $4a(3a - 2b)$

m $4bc(b + 2)$ **n** $2b(4ac - 3ed)$ **o** $2ab(1 + 2a)$

p $2(2x^2 + 3x + 4y)$ **q** $3m(2p + 3b + t)$

r $2cd(4d - 1 - 2c - 6cd)$

2 a $n(n - 1)$

b If n is odd then $n - 1$ is even and if n is even then $n - 1$ is odd.

odd × even = even

3 a length $= 4x + 3$ **b** length $= 3 - 2p$ **c** length $= t + 4$

4 $y(x + \frac{1}{2}z)$

Brackets

1 a $3a + 6$ **b** $12 + 3x$ **c** $3x - 3y$

d $x^2 + x$ **e** $7x - x^2$ **f** $x^2 - xy$

g $9a + 3a^2$ **h** $9a^2 + 3ab$ **i** $12a^3 - 9a^2b$

2 a $5x + 12$ **b** $14x + 13y$ **c** $3x^2 + 2x$

d $7x^2 + 7x$ **e** $2a^2 + 6ab$ **f** $6x^2 + 5xy + 6y^2$

g $23a^2 + 14a$ **h** $16pq - 10pr - 2qr$

3 a $3a + 6 + 9a + 9$ **b** $8b + 2 - 4b + 12$
   $= 12a + 15$ $= 4b + 14$
   $= 3(4a + 5)$ $= 4b + 16 - 2$
                          $= 4(b + 4) - 2$

4 a $y(x + 1) + x(y + 1)$ **b** $2xy + x + y$

Double brackets

1 a $x^2 + 5x + 6$ **b** $p^2 + 6p + 8$

c $a^2 + 2a - 8$ **d** $d^2 + 2d - 15$

e $p^2 - 9p + 20$ **f** $15 + 8x + x^2$

g $10 - 3y - y^2$ **h** $x^2 - 1$

i $q^2 - 100$ **j** $x^2 + 8x + 16$

k $16 + 8x + x^2$ **l** $a^2 - 4a + 4$

2 $x^2 + 2x - 15 - 2x = x^2 - 15$

3 a $(x + 2)(x - 3)$ **b** $x^2 - x - 6$

4 a $(x + 4)(x + 2)$ **b** $x^2 + 6x + 8$

Solving linear equations

1 a $x = 2$ **b** $x = \frac{1}{2}$ **c** $x = 9$ **d** $x = 5$

2 a $x = -3$ **b** $x = 3.5$ **c** $x = 4$ **d** $x = 1$

e $x = 0.2$ **f** $x = 2$ **g** $x = 1.5$

Set up and solve linear equations

1 a $10x = 180$ **b** $x = 18$ **c** 36°, 36° and 108°

2 a $\frac{n}{2} + 4 = n + 1$ **b** $n = 6$

3 a $5(x - 2) = 3(x + 4)$ **b** $x = 11$

4 a $6(2x - 1) = 8(x + 5)$ **b** $x = 11.5$

5 $x = 3$ **6** 33 years old   **7** 12 years old

8 6   **9** £2.15

Trial and improvement

1 a 5 and 6   **b** 4 and 5   **c** 4 and 5

2 $x = 4.3$ **3** $x = 5.6$

4 5.5   **5** 4.7

6 a $x(x^2 + 10) = 750$ **b** $x = 8.7$

7 $x = 2.8$

Rearranging (changing the subject of) formulae

1 a $a = x - 6$ **b** $a = y + 6$ **c** $a = 6 - z$

2 a $b = \frac{x - 6}{4}$ **b** $b = \frac{y + 3}{5}$ **c** $b = \frac{6 - z}{2}$

3 a $R = \frac{V}{I}$ **b** $b = \frac{P}{4}$ **c** $b = \frac{2A}{h}$

d $c = y - mx$ **e** $m = \frac{y - c}{x}$ **f** $t = \frac{v - u}{a}$

g $\pi = \frac{A}{r^2}$ **h** $C = \frac{5(F - 32)}{9}$ **i** $a = \frac{A}{\pi b}$

4 a $r = \sqrt{\frac{A}{\pi}}$ **b** $r = \sqrt{\frac{V}{\pi h}}$ **c** $r = \sqrt[3]{\frac{5V}{4p}}$

The nth term

1 a $5n$ **b** $2n$ **c** $50n$ **d** $20n$

2 a $5n + 1$ **b** $2n + 1$ **c** $4n - 1$

d $5n - 4$ **e** $3n - 7$ **f** $10n - 90$

3 a $13 - 3n$ **b** $24 - 4n$ **c** $107 - 10n$

d $-5n$ **e** $3 - 3n$ **f** $11 - 10n$

4 a $\frac{2n + 1}{5n - 1}$ **b** $\frac{n}{3n - 1}$ **c** $\frac{4n - 1}{107 - 10n}$ **d** $\frac{3 - 4n}{n + 5}$

5 $4n - 2$

6 a $3n + 2$ **b** 50 tables **7 a** 2^n **b** 10^n

8 a 7, 12, 17, 22 **b** 502

9 a 1, 3, 6, 10 **b** 5050

10 a 2, 6, 12, 20 **b** 10 100

11 a 0, 5, 14, 27 **b** 19 899

Inequalities

1 a $x < 5$ **b** $x \geq 14$ **c** $x \leq 7$

d $x < 4$ **e** $x \geq 1$ **f** $x \leq -2$

g $x > 30$ **h** $x \geq 8$ **i** $x < -2$

2 a $x = 5$ **b** $x = 13$ **c** $x = -5$

d $x = 4$ **e** $x = -6$ **f** $x = 2$

3 a $x = 9$ **b** $x = 13$

4 a $x \geq 1$ **b** $x < 3$ **c** $0 < x \leq 4$ **d** $-3 \leq x \leq 1$

5 a 1,2 **b** 1,2 **c** 1

6 a 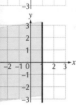 **b**

c **d**

e **f**

7 $x \leq 1.5$

8 a  **b**

c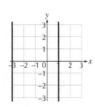

Real-life graphs

1 a i 12:00 **ii** 13:00 **iii** 15:00

b i 20 mph **ii** 40 mph

2 a 6 km/h **b** 0.6 m/min **c** 0.8 km/h **d** 1.4 km/min

3 a 40 km/h 5 km/h 0 km/h 30 km/h

b A to B – it is the steepest section.

4 Rob keeps a steady pace throughout to win the race. Darren sets off more quickly, but slows down and is overtaken. He then speeds up again but can't catch Rob.

5 a A = 4 C = 1 D = 5 E = 3 F = 2

b

Gradient and intercept

1 A = $\frac{1}{2}$ B = 1 C = 2 D = 2 E = -2 F = $-\frac{1}{2}$

2 a 4 **b** 1 **c** $-\frac{1}{2}$ **d** -2

3 a line a **b** it has the biggest gradient

4 a 4 **b** 3 **c** 2 **d** 0

5 a D **b** A and E, B and C, F and G, H and I

c I **d** D **e** 4

Drawing graphs – linear

1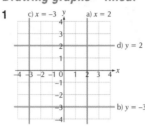

2 a

x	-2	-1	0	1	2
y	-5	-3	-1	1	3

b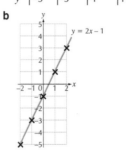

c $x = 0.5$

3 a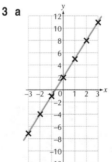

b $x = -2$

4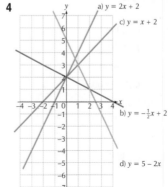

Answers

Drawing graphs – quadratic

1 a

x	–4	–3	–2	–1	0	1	2	3
y	18	11	6	3	2	3	6	11

b

$y = x^2 + 2$

c $y = 4.25$

2 a

x	–2	–1	0	1	2	3
y	15	5	–1	–3	–1	5

b

c i look at points where curve crosses x-axis

ii 2.22

ANSWERS TO GEOMETRY AND MEASURES

Circles – circumference

1 a 31.4 cm   **b** 11.3 mm   **c** 596.9 m

    d 62.8 cm   **e** 22.6 mm   **f** 1193.8 m

2 a 20π mm   **b** 62π cm   **c** 7.6π m

3 a 204.2 cm   **b** 244   **c** 1 714 005

4 113 097 cm ÷ 125.7 cm = 899 revolutions

5 a 25.7 cm   **b** 21.4 cm

6 a $(5\pi + 20)$ cm   **b** $(10\pi + 30)$ cm

7 95.5 cm   **8** 0.16 m   **9** 99.4 cm

Circles – area

1 a 78.5 cm²   **b** 10.2 mm²   **c** 28 352.9 m²

    d 314.2 cm²   **e** 40.7 mm²   **f** 113 411.5 m²

2 a 100π mm²   **b** 961π cm²   **c** 14.44π m²

3 a 78.54 cm²   **b** 257.08 cm²

4 a 21.46 cm²   **b** 85.84 cm²   **c** 23.18 cm²

5 a 128π cm²   **b** 64π cm²   **c** 32π cm²

    d 16π cm²

Prisms – surface area

1 190 cm²   **2** 72.16 cm²   **3** 84 cm²   **4** 211.2 cm²

5 240 cm²

6 272 cm²

7 184.8 cm²

8 727.8 cm²   **9** 664 cm²   **10** 296 cm²

11 10

Prisms – volume

1 150 cm³   **2** 96 cm³   **3** 261 cm³

4 192 cm³   **5** 860 cm³   **6** 288 cm³

7 1100 cm³   **8** 37 699 cm³   **9** 1590.43 cm³

10 Cuboid tin (**A**):   8.5 × 15.5 × 15.5 = 2042.125 cm³

     Trapezoid tin (**B**): $\frac{1}{2}(14 + 23) \times 8.5 \times 14 = 2201.5$ cm³

     Cylindrical tin (**C**): $\pi \times 9^2 \times 8.5 = 2162.99$ cm³

     Trapezoid tin has the greatest volume

Density

1 a 45 cm³   **b** 189 g   **c** 0.8 g/cm³

    d 850 g   **e** 600 cm³   **f** 0.9 g/cm³ or 900 kg/m³

2 33 180 kg   **3** 500 cm³   **4** 1.6 g/cm³

5 86.4 g   **6** £344 968.20

Pythagoras

1 a 13   **b** 7.1   **c** 9.9

2 a 6   **b** 6.6   **c** 7.2

3 14.1 cm   **4** 131.4 m   **5** 3.3 m

6 5.1 cm   **7 a** 5 units   **b** 8 units

8 16.40

Regular polygons

1 a 108°   **b** 72°

2 A hexagon can be divided into 4 triangles. The angle sum of a triangle is 180° and 4 × 180° = 720°.

3 8

4 External angles = 180 – 160 = 20° so number of sides = 360 ÷ 20 = 18

5 $a = 120°$, $b = 30°$, $c = 90°$

6 64°

Translation

1 a

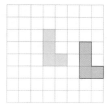

b

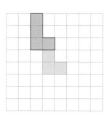

2 a 1 square right, 4 squares up

b 4 squares left, 3 squares up

c 1 squares left, 4 squares down

3 a $\begin{bmatrix} 4 \\ 0 \end{bmatrix}$ **b** $\begin{bmatrix} 0 \\ 4 \end{bmatrix}$

c $\begin{bmatrix} 2 \\ -3 \end{bmatrix}$ **d** $\begin{bmatrix} -4 \\ 4 \end{bmatrix}$

e $\begin{bmatrix} -4 \\ -4 \end{bmatrix}$ **f** $\begin{bmatrix} -2 \\ 3 \end{bmatrix}$

4 $\begin{bmatrix} 6 \\ 4 \end{bmatrix}$

Rotation

1 a 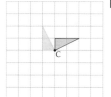 **b**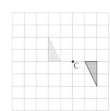

2 a
b
c

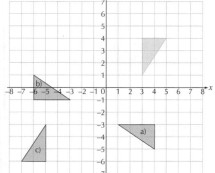

3

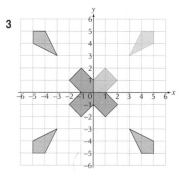

Reflection

1

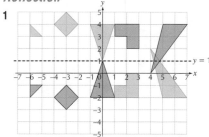

2

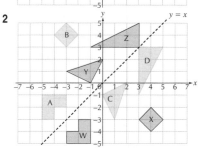

3 a Reflection in the line $x = -3$

b Reflection in the line $y = -1$

c Reflection in the line $y = -x$

d Reflection in the line $y = x$

Enlargement

1

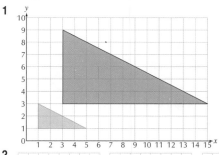

2

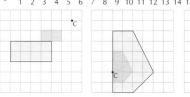

Answers

3

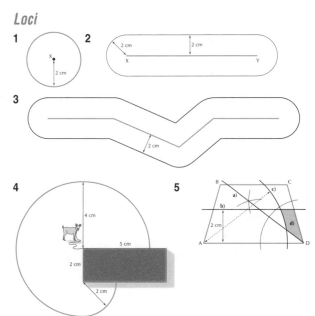

scale factor ⅓

scale factor ½

scale factor ¼

Constructions

Check that all constructions are accurate and that construction lines are shown.

Loci

1

2

2 cm

2 cm

X   Y

3

2 cm

4

4 cm

5 cm

2 cm

2 cm

5

B   a)   c)   C

b)

2 cm   d)

A   D

Dimensional analysis

1 $2ab + 2ab + 2ab$ represents total A

abc represents total V

$4a + 4b + 4c$ represents total L

2 a A   **b** l   **c** L   **d** V   **e** X   **f** L

g A   **h** L   **i** X   **j** V   **k** l   **l** V

3 a l   **b** N   **c** V   **d** A   **e** N   **f** A

Midpoints

1 AB = (3, 5)   CD = (–0.5, 1)   EF = (–2, 3)

GH = (3, 4)   IJ = (1, 0)   KL = (0.5, 0.5)

2 a (2, 1)   **b** (–1, 2)   **c** (0, 2)   **d** (1.5, 2.5)

3 a (2, 2.5)   **b** (2, 2.5)

4 a (4, 8)   **b** (3, 6)   **c** (5, 2)

Mean, median, mode and range

1 a 6   **b** 7   **c** 7   **d** 5.9

2 a £29 000   **b** £18 000   **c** £21 000   **d** £21 500

e £29 000   **f** £20 000   **g** £23 000   **h** £23 500

3 a 3   **b** no mode   **c** 4   **d** 3.5

4 a 6   **b** 6   **c** 6   **d** 5.7

5 a 34   **b** 48   **c** 49   **d** 49

6 a 23 cm   **b** 17 cm   **c** 26 cm   **d** 26.9 cm

7 a $5 < x \leq 10$ **b** 9.4

8 a $20 < x \leq 30$ **b** 27

c We do not know the highest or lowest values.

9 6.8   **10** 84 kg   **11** 78.8 kg

12 a 124   **b** 2.54   **c** Year 8, with a mean of 2.9

13 2.35

Correlation

1 a

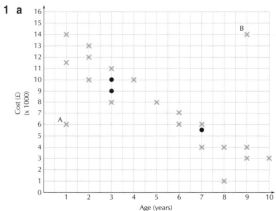

b Yes, graph shows negative correlation.

c i A cheaper make/model of car

ii A more expensive make/model of car

d The trend can't continue indefinitely as the car would be worth a negative amount. Eventually the car will be worth a minimum or 'scrap' value.

2 a Yes, the graph shows positive correlation.

b Low attendance but high mark

c Strong positive correlation

d

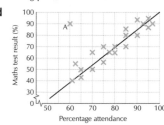

e i 75–82%   **ii** 70–75%

f This value lies outside the range of the given data.

3 a&b

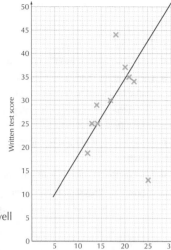

c 17–19

d Dom – position lies well below line of best fit.

e Positive correlation

Retail price index

1

2005	2006	2007	2008	2009
18.7	17.9	20.2	24.8	22.9

2 $1144   **3** 560 000

4 a 69p per $ **b** 88

Questionnaires

1 a For example, it is a leading question and it does not give the option to disagree.

b For example, the question is clear and all possible answers are covered.

2 a Check that there are a sensible number of categories, that they do not overlap, and that all possible answers are covered.

b Check that there are a sensible number of categories given and the option 'Other' for anything that hasn't been covered.

c Students do not have an equal chance of being chosen.

d Method 2 as it is better to make a random selection. If he used method 1, it is likely that the first 50 students he sees at break will include large groups from the same Year group.

Probability

1 a 0.3, 0.12, 0.19, 0.184, 0.163   **b** $\frac{1}{6}$ **c** 1000

2 500

3 a 100   **b** 200   **c** 250   **d** 0

4 Janet, as she has carried out more trials and her frequencies are all close to 50.

5 a $\frac{1}{2}$ **b** $\frac{1}{2}$

c A card cannot be both red and black.

d A card must be either red or black.

6 a HHH, HHT, HTH, HTT, THH, THT, TTH, TTT

b $\frac{1}{8}$ **c** $\frac{3}{8}$

7 $\frac{1}{36}$

8 a

	1	2	3	4
1	2	3	4	5
2	3	4	5	6
3	4	5	6	7
4	5	6	7	8

b $\frac{3}{16}$

9 $\frac{1}{2}$ **10 a** 0.27   **b** 850   **c** 357

ANSWERS TO REVISION UNITS

Paper 1

1 a $\frac{100}{5} = 20$ **b** $\frac{450}{10} = 45$

2 $3n + 1$

3 a $2(n + 4)$ double any number and get an even number

b 9th term   **c** $\frac{4n - 3}{38 - 3n}$

4 For example, $n = 1$ gives $n^2 + 2 = 3$.

5 a 30 cm, 31 cm, 32 cm, 33 cm, 34 cm

b

```
27  28  29  30  31  32  33  34  35  36  37
```

6 $a < 5$ **7 a** $2^2 \times 3^3 \times 5$ **b** 45 cm

8 a 5

b No, $6^{15} \div 6^5 = 6^{10}$. The rule is to subtract the indices.

9 2008: 1000 × £10 = £10 000
2009: 1200 × £8 = £9 600

No, the florist takes less money in 2009.
OR 1.2 x 0.8 = 0.96, 96% is less so the florist takes less money

10 22 packs × £15 = £330

11 a

x	–3	–2	–1	0	1	2	3
y	17	10	5	2	1	2	5

b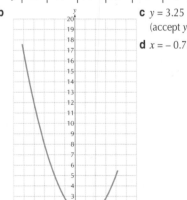

c $y = 3.25$
(accept $y = 3$ to 3.5)

d $x = -0.75$ and $x = 2.75$

Answers

12 a i

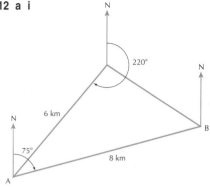

ii approx 4.6(±0.2) km on a bearing of 303°(±5°)

b

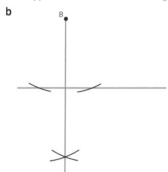

13 a,c&d

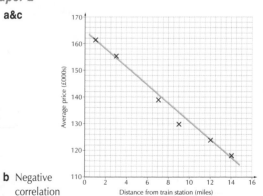

b $y = -1$

Paper 2

1 a&c

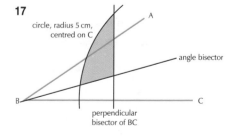

b Negative correlation

d 4.5 miles approx (answer may vary slightly depending on line of best fit)

e This distance lies outside the given information.

2 a 2.3 minutes (or 2 minutes 18 seconds)

b We don't have each shopper's individual time.

c $1 < m \le 2$

d Yes, 52% chose in less than 2 minutes.

3 a For example: There is no option for less than 2 hours or more than 8 hours or the categories overlap.

b Check that the answer covers all possibilities and has no overlapping categories.

c For example: It is a leading question or there is no option to disagree.

d For example: Which is your favourite games consul?

X-box   Playstation   Wii   Other

4 a $\frac{13}{50}$ **b** Don't know the individual values spent.

c 360

5 £2027.50

6 No : 1560 − 975 = 585 and 585 ÷ 975 × 100 = 60%

7 Yes: percentage increase is 5% and 546 000 × 1.05%
= 696 850 ≈ 700 000

8 Ratio = 3 : 1   Farrouq receives £750, Mahmoud receives £250, difference = £500

In 4 years' time ratio is 5 : 3

Farrouq receives £625, Mahmoud receives £375, difference = £250

9 a £54   **b** 1100 units

10 a $x = 2.7$ **b** $3x^2 - 4x$ **c** $9d + 14$

11 $x = 3\,cm$

12 a 45°   **b** 135°   **c** 22.5°

13 Hypotenuse length = 16 m
Total length = 38.4 m ÷ 0.4 = 96 blocks   96 × £1.50 = £144

14 Area of cross-section = $\frac{1}{2} \times 3 \times 4 = 6\,cm^2$
$x = 54 \div 6 = 9\,cm$

15 a Perimeter of semicircle = (20π ÷ 2) + 20 = 51.4 cm
Side length of square = 51.4 ÷ 4 = 12.85 cm

b Perimeter of circle = π × 30 = 94.2 cm
Side length of square = 94.2 ÷ 4 = 23.55 cm
Area of circle = π × 15² = 706.9 cm²
Area of square = 23.55² = 554.6 cm²
The circle has the largest area

16 $x = 5.1\,cm$

17

circle, radius 5 cm, centred on C

angle bisector

perpendicular bisector of BC